From Terrain to Brain

From Terrain to Brain

Forays into the Many Sciences of Wine

ERIKA SZYMANSKI

OXFORD
UNIVERSITY PRESS

Oxford University Press is a department of the University of Oxford. It furthers
the University's objective of excellence in research, scholarship, and education
by publishing worldwide. Oxford is a registered trade mark of Oxford University
Press in the UK and certain other countries.

Published in the United States of America by Oxford University Press
198 Madison Avenue, New York, NY 10016, United States of America.

CIP data is on file at the Library of Congress
ISBN 978-0-19-764031-9

DOI: 10.1093/oso/9780197640319.001.0001

Printed by Sheridan Books, Inc., United States of America

The images preceding each chapter title page and the icons used throughout the book were designed and created by Sumanma Wadhwa.

Contents

Introduction

The world of wine is vast and crisscrossed by innumerable connecting paths. Wine is a subject for physics and chemistry and biology, for plant and microbial genetics, fluid dynamics, color science, sensory science, soil science, and nearly every other natural science you might think to name. It's studied by researchers in applied scientific domains such as medicine and horticulture. It's significant in history, psychology, sociology, anthropology, geography, literature, rhetoric, economics, and marketing. Each studies wine through its own lens. In every case, looking through one lens invokes others, so that ways of knowing anastomose around wine as a social, cultural, and scientific phenomenon.

The many sciences of wine, in short, are messy and multiple. That's a good thing.

Science—or sciences, since there are many of them—is about building pictures of what things are and how they work. The more detailed our pictures become, the more we can see; the more we can see, the more we can enjoy. Unfortunately, wine science is often presented as a rather settled list of facts to memorize. Memorizing facts is tremendously useful for acing exams or showing off at parties, but otherwise maybe less useful for enabling enjoyment. And talking about facts is inclined to give the impression that those facts are the singular, authoritative way of describing what things *are*, rather than one way to make sense of them. Science begins as a mode of organized curiosity, a practical enterprise of figuring things out, following one of multiple potential paths for doing so. Turning science into lists of facts thereafter is a choice.

This book contains plenty of facts, well seasoned with hypotheses and speculations that might grow into facts if researchers keep researching them. But they're not really the point. The point is that asking questions about wine, trying to get at how wine works, can't help but pull together ways of thinking about the world that often otherwise seem to be separate.

The next two sections of this introduction lay out some theory that scaffolds how I think about what wine sciences are and what they're good for. If you don't care about the theory, you might want to skip to the OBLIGATORY WARNING and HOW TO USE THIS BOOK at the end of this introduction, or skip straight ahead to any of the chapters.

What Is Wine Science, and What Is It Good For?

What is wine science, and what is it good for? I ask myself those questions often. As is generally true of questions worth repeating, the answers are context-sensitive; they depend on why you're asking. But the underlying argument of this book is that for most people, in most cases, wine science constitutes a set of strategies for seeing. That's the big picture. To make that picture more detailed, I need to lay out three principles about how scientific approaches to wine work.

"The science of wine" doesn't exist. There's no one systematic way to make sense of wine. There are many. Wine can be studied and described as a non-ideal solution or a nearly ideal fluid, a treasure trove of biologically active organic compounds, a microbial environment, a nutritious food, a toxic drug, a value-added agricultural product, a means of preserving fruit, a historical artifact, a local tradition, an economic entity, bottled poetry, and so on.[1] Those approaches aren't mutually exclusive. They intersect, but beginning with any one of them leads to asking different kinds of questions about how wine works, each useful in its own ways.

Wine science is not a set of facts. Science is a set of processes, and the knowledge built through those processes changes over time. In other words, facts are unstable; they're subject to change. Some are more unstable than others. Scientific research can't help but involve incremental fits and starts, including some wrong turns and missteps that only become recognizable as such further down the road.

---

[1] Oenology and viticulture, the studies of winemaking and grape-growing, respectively, are themselves problem-oriented fields that draw on multiple approaches.

Misunderstanding science as a set of facts is the root of all kinds of problems. People who aren't scientists are sometimes surprised, dismayed, even angry when "the science" changes. People who *are* scientists are sometimes surprised, dismayed, even angry that everyone else doesn't "do what the science says." Neither leaves much room for other folks doing things with scientific knowledge other than what *they* do—a restriction totally at odds with the idiosyncratically place-based, deliciously diverse, occasionally artistic world of wine.

Wine science isn't a set of instructions about how to make wine. It follows from the first two points that science is not a prescription for How Things Should Be Done. All research arises in specific physical and cultural locations, moments in history, motivations for doing the work, and so on. Findings from a drought-plagued Californian Pinot Noir vineyard won't necessarily transfer to a drippier one in Oregon or Burgundy; they must be interpreted in light of local conditions. Moreover, science is *one set* of ways of knowing about the world, not the only or universally best way. Ideally, scientific findings about intensifying color in a particular Californian pinot noir work together with what can be learned from walking vineyards and watching ferments, wherever you are.[2]

Science isn't good for everything. Mastery of the phenolic chemistry behind wine astringency and color will never replace the experience of drinking pinot noir, any more than knowing about the chemical composition of an artist's paints replaces beholding a painting. Most of all, because research is always grounded in values and perspectives about what is important to know and why, you can hold science in high esteem without agreeing with the idea of "better wine" baked into any given part of it. The wine industry is replete with examples of practitioners reading scientific research, integrating it with other things they know, and coming up with something that researchers didn't necessarily expect or intend—though the scientific establishment isn't always thrilled about that.[3]

---

[2] In this book, I follow the convention of capitalizing grape varieties, but not capitalizing the varietal names of wines made from them.

[3] In some ways, wine is like any other agricultural industry, but in this way, it's not. Everyone who grows commodity corn likely shares an idea about what constitutes good corn. Ideas about good wine vary far more widely, so individual and institutional values and goals are far more consequential.

In the academic and practical field known as "research utilization" or "technology transfer," the goal of communicating research is usually "adoption"—that is, convincing your target audience to implement research findings as intended by the organization doing the communicating. Full disclosure: I wrote a 443-page doctoral thesis about (among other things) why that approach is theoretically insupportable and practically misguided, especially in the wine industry.[4] The most important thing to say from all of that work is this: winemakers and wine-growers use scientific research in innumerable ways that don't necessarily align with the expectations of scientific institutions because those winemakers may have knowledge, values, and priorities that differ from the institutions'. Vanishingly few winemakers are "anti-science."[5] Most take pains to learn about new research. Most account for research findings, alongside practical experience and local convention, when they have a decision to make or a problem to solve. None of that means that they "do what the science says," not least because research rarely spits out instructions—though universities or national wine programs and other institutions sometimes make research findings *into* instructions in the context of institutional goals and values. Research use, in short, needs to be interpreted in light of what any one "user" might be trying to achieve. For the same reasons, writing and learning about wine science isn't about any one end, either.

I came away from my PhD with the sense that people who most strongly self-described as "following the science" rarely made wines that I was excited to drink. I think that's because following the science often meant paint-by-numbers winemaking that prioritized "correct" or "safe" choices, at odds with centering the idiosyncrasies of local context or a personal vision of good taste, and sometimes at odds with prioritizing sensory experience over numbers. Number-based guidelines about the "correct" way to do things rarely apply universally across contexts and goals; sensory experience, meanwhile,

---

[4] Szymanski, "Through the Grapevine."
[5] All wine regions have their quirks. I conducted my PhD research in New Zealand and Washington State, and I have good reasons to believe that those two places aren't outliers in this respect, but they certainly have their own distinct cultures.

locates you in a specific *here* and *now*. That's not to say that wines made through applying contemporary science and technology are always boring—far from it—but that prioritizing scientific universals (or would-be universals) trends toward more uniform, less interesting wines.

Of course, "interesting" isn't everyone's priority. You may care more about consistency, or favorite flavors, or supporting particular people or places. That's all fine too. Indeed, that's the point; the same data can be used in myriad ways depending on what you're trying to do.

So much for what wine science isn't. Defining what the many sciences of wine *are* is tough because that's contextual too; there are many good answers, not one right or best one. For this reason, I've built chapters around topics that aren't owned by one field but that instead integrate multiple perspectives—that take forays through wine worlds in the spirit of a research mode called science and technology studies.

A Note on Forays

The "foray" in the subtitle of this book is an appreciative nod to Jakob von Uexküll's *A Foray into the World of Animals and Humans,* initially published in German in 1934. Von Uexküll was a theoretical biologist and is now often described as an early intellectual innovator in cybernetics, the theory of information, control, and communication that temporarily threatened to dominate the English-speaking world's approach to science and technology after World War II. His *Foray* has earned the right to be remembered a century later for the idea that each creature inhabits its own world—its *umwelt* or environment—constituted by that creature's particular abilities to perceive and act. The flavor chapter has more to say about that. ♪

The forays in this book make little trips out into worlds of wine science, through territories that extend from terrain to brain, from intersections that constitute *place* to cognitive experiences. I hope that these outings will help expand your own *Umwelt* by making connections that you might not otherwise have perceived.

A Note on Science and Technology Studies

Contemporary science excels at learning about things by disassembling them. It's less well-suited to putting them back together. Scientific research often isolates one aspect of how something functions by standardizing its surrounding environment, so that the environment effectively drops out of the experiment. 🖐 Having made context invisible for this purpose, recalling that scientifically demonstrated phenomena function in a slew of varied contexts out in the *rest* of the world can be difficult. Science and technology studies or STS, the interdisciplinary arena in which I primarily live my academic life, takes (re)locating science and technology in context as its central objective. That's what I'm trying to do here.

Obligatory Warning

I've tried to make this book accessible to anyone who might realistically want to pick it up. That means that I've taken for granted that you, reader, find wine interesting and know a bit about it, but not that you know anything out of the ordinary about sciences. I've therefore sacrificed lots of detail in the service of keeping a bit. I've tried to tell one story at a time without stumbling down too many others, even though wine is a morass of tangled paths. I've flattened nuance and no doubt made some errors, and even things that are generally accepted as true while I'm writing may cease to be so by the time you're reading. If you're seeking a comprehensive treatment of any one scientific concern, please don't look here, because that's not what I'm trying to do.

Frankly, I'm not sure what being comprehensive would mean. Where do intersections of wine and sciences begin and end? Even if that question had an answer, any attempt to "cover" all of them would be quite the superficial blanket, and that's no fun. Instead, I've taken trips through topics, diving more or less deeply in an effort to catch some illumination. The result is that all manner of obviously important things are left out. Tannins get short shrift. So do vine diseases, soil structure, and effervescence. I've barely touched upon the biochemistry, microbiology, human physiology, and neuroscience that contribute to how we experience wine flavor. I've ignored the entire field of color science and what it has to say, in conversation with chemistry, history, and psychology, about how wine color develops and why it matters. In every case, where I've said something, there's more to be said. That's a limitation of the approach that I've chosen, but also part of the point I'm trying to make: that there are no complete and final words on the many sciences of wine.

Similarly, the examples I've chosen aren't comprehensive geographically or culturally. They often extend from research I've conducted, primarily in the western United States, Australia, and New Zealand. As you might expect, their generalizability varies. Geography is part of context, and while scientific research has studied and shaped wine the world over, it hasn't done so the same way everywhere. In particular, wineries in some parts of Europe can be legally constrained against experimentation in ways that those elsewhere rarely are.

The wine world is pretty WEIRD: Western, educated, industrialized, rich, and democratic.[6] It is, in other words, unlike the world that most humans inhabit today, and even more niche in any kind of historical perspective. I use terms like "we" and "social context" as shortcuts to reference this present peculiar little subset of humanity that drinks wine and tends to read popular science books in English, because constantly typing out the longhand version is hopelessly cumbersome. I hope that the limitations of those shortcuts become more obvious over time.

How to Use This Book

Each of the following chapters is a short trip or foray through a topic that synthesizes scientific approaches to what wine is and why it matters. Every chapter can be read independently in any order, and I encourage you to begin wherever you please, though the book is organized very roughly along a line from the beginnings of winemaking

---

[6] Wine research, more than many other scientific specialties, has also remained mostly white and male—a perspectival imbalance that can't be avoided especially for science prior to the twenty-first century.

(terrain) to the ends of drinking it (brain). Small icons in the text indicate significant intersections among chapters. In general, I've tried to explain everything you need to know to make sense of a chapter in that chapter. But were I to follow that notion to its logical ends, every chapter would need to be an omnibus of every other chapter, so fuller explanations of adjacent points are sometimes located elsewhere. I use footnotes for details that don't quite fit but that I want to acknowledge, and for references.[7]

I've enjoyed writing this book. I hope that you enjoy reading it, no matter what you might be trying to achieve.

---

[7] You may find yourself wanting to follow up on something I've cited. Many of my references are academic journal articles that, at the time I'm writing, live on the Internet behind stupidly high paywalls. If you want to read one of these (though most academic articles aren't exactly accessible reading independent of the journal's payment policies), contact your friendly local university librarian, a friend who works or studies at a research institution, or me. We can help.

1
Geography 🍷

Geographical indications sound as though they should be about geography. They are, but only when "geography" is defined broadly enough to encompass the history and storytelling folded into how map lines are drawn. Indicating anything at all about geography is ultimately about assigning names to places—how places are *made* and not just found, about storytelling and not just regional distinctiveness. That built-in multidimensionality means that we can never simply ask whether a wine-growing region warrants its own name. Instead, the question has to be: does someone want to make that region its own place, and are their interests more influential than those of anyone who might disagree? Whether a wine-growing region is one place, three distinct places, or no place special has everything to do with the way you look at it, across space, and across time.

Hilgard and the Origins of Californian Wine Regions

In 1880, Dr. Eugene Hilgard asked the California state legislature for $4,000 to study viticulture at the University of California, and ended up with $3,000 annually for a new wine research program.[1] At the time, Hilgard was *the* professor of agriculture for the entire university and, thus, the state. (Last I checked, the University of California Davis's College of Agricultural and Environmental Sciences had 381 faculty and over 1,000 graduate students. Don't ask how many agriculture professors are serving the whole state of California.)

In wine science terms, 1880 is more than halfway to ancient. Pasteur had only demonstrated a firm association between alcoholic

---

[1] Amerine, "Hilgard and California Viticulture."

fermentation and the critters now known as *Saccharomyces cerevisiae* in 1857 (though it should be acknowledged that some of his predecessors had similar suspicions before then). Wine itself has been around for some 7,000 years or so and humans have surely been trying to make it better for nearly as long, but oenology as we know it is the product of the past century or so.

In 1880, shiny analytical chemistry instruments were fewer and simpler. Chemists couldn't wield mass spectrophotometry or liquid chromatography to study specific wine molecules. The whole field of modern genetics, with its abilities to engineer yeast and decipher variety lineages and understand vine diseases, didn't exist. No plant pathologist or computer scientist was mining big data for patterns in close-up images of grape leaves. So, what did a wine scientist *do* back then?

A wine scientist made wine, mostly, and evaluated whether anyone would want to drink it. Hilgard—who initially trained in geology and chemistry, then became a professor of agriculture without ever specializing exclusively in wine—took it upon himself to systematically investigate which grape varieties were worth growing for Californian wine, where each could reasonably be planted, which winemaking techniques were suitable for getting the best out of them, and how their end products could be expected to taste.[2] (This was a gargantuan task.) California could grow great grapes; that was abundantly clear. But great wines came from Europe. So the question became: what did California need to do to make European-style wines? To Hilgard, figuring out what kinds of grapes to plant where was at the heart of answering it.

Hilgard's grand project to map Californian viticulture was bizarrely overambitious. It may also have headed off a lot of boring wines. By 1925, his successors could conclusively recommend that growers avoid "Bambino bianca, Clairette blanche, Feher Szagos, Green Hungarian, Hibron blanc, Hungarian Millenium, Kleinberger, Malmsey, Marsanne, Mathiaszyne, Mourisco branco, Muscat Pantellana, Muscat Saint Laurent, Nasa Veltliner, Nicolas Horthy, Palaverga, Pavai, Roussette,

---

[2] Amerine, "Hilgard and California Viticulture."

Saint Emilion, Sauvignon vert, Selection Carriere, Steinschiller, Terret, Vermentino Favorita, and Vernaccia bianco"—and those were just the whites deemed "so devoid of merit that they cannot be recommended anywhere in California under present conditions."[3] Their failings were varied: disease-prone, insufficient acid or sugar, inconvenient ripening habits, or just plain dull. "Petite verdot" (the terminal "e" on "petite" was an error) earned a spot on the reds-not-recommended list for producing only "a standard though slow-maturing and rough wine," producing a meager crop, and sunburning too easily. Petit verdot excepted, their judgments have held remarkably steady, considering that they produced their test wines by the brutal method of pulverizing the grapes between rollers and yanking the stems out by hand.

Hilgard's system for classifying grape varieties relied on where those varieties found their traditional European homes, and where similar conditions might exist in California. He ended up dividing the entire state into two viticultural regions: coastal and non-coastal. By the 1920s, when his successors were finally getting a handle on his initial, insanity-inducing project, they realized that his original distinction wasn't exactly optimal. In its place, they outlined five "climatic zones," with a footnote that future researchers might, somehow, find a way to make more. In addition, instead of strictly referencing European styles, they classified wines by their quality characteristics: reds and whites with specific varietal character such as zinfandel or chardonnay, those without such character but still of "general bottling quality," and those not worth bottling but presumably suitable for plonk by the barrel.

These heuristics may look rudimentary and even a bit silly from today's vantage, but it's a fallacy of popular history that our predecessors were stupider than us. They weren't. (Wine lovers are probably less inclined toward that mistake than most, since rhapsodizing over the elegant produce of our forefathers might conflict with deriding them as ignoramuses.) Such insensitive (to contemporary eyes) designations no doubt made sense at the time as steps toward raising Californian wines—largely *not* elegant produce in 1920—against French and Italian benchmarks.

---

[3] Amerine and Winkler, "Composition and Quality," 606.

Today, California is formally subdivided into 139 American Viticultural Areas (AVAs), from Napa Valley, instituted in 1981, to the Petaluma Gap, instituted in 2018. If you find someone who still isn't convinced that California's wines can hold their own against bottles from the other side of the Atlantic, put a star on the calendar; they're a rare find. Californian wines have come a long way, unquestionably helped by putting a finer point on what grows best where.

Except that those 139 AVAs may not be veritably useful to wine drinkers, or to marketers, or even to researchers who live to subdivide and define things. And Hilgard was further ahead than I've painted him in realizing that Californian wines couldn't and shouldn't aim for European mimicry. In his report for the State Viticultural Commission in 1885, his conclusion was this:

> [If] the wines of California must in the main seek their market outside of the State, and must therefore be adapted to shipment to long distances; then it follows that, if we adopt the wine-making processes of southern France, Portugal, and Italy, we must adopt the all but universal practice of fortifying export wines. If, on the contrary, we wish, in our climate, to produce also wines similar to those of Bordeaux and northward to the Moselle, we must of necessity so vary our practice that with grapes of a more or less southern character we may nevertheless be able to impart the characteristics of the cooler climates to our products. To this end we must distinctly deviate, in some respects, from the exact practice of either the southern or northern region of Europe.

Copying Europe wasn't suited to the Californian climate—not just the weather, but the *business* climate. For example, a major concern for the Californian grape industry in the early 1920s was railroad capacity. Grapes were literally rotting in the vineyard because too few railroad cars were available to cart them cross-country for winemaking in eastern states. Another project undertaken by the state's viticulturists therefore involved trialing preservatives to keep grapes fresh longer in transit. They concluded that sulfur dioxide worked fairly well when applied either by wafting it through the rail car or by submerging grapes in a solution; ⊗ boric acid, formic acid, and formaldehyde, on the other hand, didn't. Thank goodness.

Some of these founding-fathers' wine-making concerns are no longer relevant, but some of them are. Hilgard saw not letting grapes hang to "'dead ripe' as is usually done" as a probable solution to the "excessive headiness for which California wines are thus far noted." He complained that vintners ignored that suggestion because, he thought, they remained trapped in a European mindset that equated more hang-time with more quality. Davis scientists noted that this was still an issue in 1944, and critics have echoed them ever since. More recently, though, in light of rising temperatures and interest in lowering alcohols, Californian viticultural innovations are more likely to be about growing ripe grapes with fewer sugars, not more. ⤳.

New(er) techniques, more mature vines, generations of experience, and a near total lack of concern about trains all mean that Hilgard's grand mapping project is unquestionably out of date, even if it could ever have been called finished in the first place. On the contrary, wine researchers are still working on it: what grows where (and how), what winemaking techniques work well (and why), and what flavors to expect (and what consumers think of them). But now, California and other wine regions are staffed with marketing researchers, economists, and soil scientists, as well as geologists-turned-agriculture professors, whose work collectively tells us that legislated place-based designations are more than simple recognitions of the physical characteristics of a bit of land. *Places* are things we make, not things we find (see the PLACE box).

Making Places

The core idea of *terroir* is that by virtue of being produced in a particular area, we can expect a wine to have specific quality attributes. ✿ Those expectations invoke the traditions of how wine is made in a given place and not just the soil or climate; *terroir* is a product of how humans work with environments and the animal, vegetable, and microbial life with which we coinhabit them. Yet the endgame remains: *because of place A, therefore wine B*. Whether that conclusion

PLACE

Geographers debate definitions of place professionally, though (as with most areas of research) what that means has changed over time. The earliest documented use of the word comes courtesy of Eratosthenes toward the end of the third century BC, when he became known for devising clever strategies for measuring the circumference of the earth and other Terran vital statistics. (He was better known, and better remembered, for heading the famed and now long-lamented Library of Alexandria at the ego-crushing age of thirty.) That was enough to cement him as a "father of geography" in the historical record, though it evidently didn't impress his contemporaries, who called him "beta" because he gave his attention to so many different subjects that he was always only second-best in any of them. Today, as part scientist, part geographer, part poet, and all-round overactive learner, he'd be called a Renaissance man. He probably would have made a good winemaker. He probably also would have made a good contemporary geographer, in a field that has moved from measuring the globe to making sense of how humans and other creatures interact to construct environments.

is justified is an entirely separate question from whether place names on wine labels work as quality guarantees for the people buying the bottles.

Wine economics research is inconclusive on that point. Some studies show that vineyard prices—a typical if questionable proxy for wine quality—correlate with the locale where grapes are grown. Some don't. More telling is that some such studies *do* show a relationship between vineyard price and geographic indication—that is, with the *name* associated with the locale. Vineyards in the region legally designated as Champagne were worth €1 million per hectare in 2017, according to the French agricultural statistics ministry; vineyards next door but just outside those borders sold for a meager 1.3 percent of that. In 2007, vineyards in the Dundee Hills subregion of the Willamette Valley garnered a premium of $7,163 per acre over vineyards entitled to the

Willamette Valley name alone.[4] Obviously, some stakeholders will want geographic indications to be more expansive than others, while a larger place is likely to have a harder time describing its unique value proposition. That tension has driven the designation of subplaces in the form of recognized viticultural areas such as the Dundee Hills, contained entirely within a larger recognized viticultural area.

But what does "place" mean? When American Viticultural Areas or AVAs were first introduced by the Bureau of Alcohol, Tobacco, Firearms and Explosives in 1978, a potential AVA was defined as a state or several contiguous states, a county or several contiguous counties, *or* "a viticultural area" that had become recognized by its own name.[5] In that last case, precise boundaries become a matter of local negotiation. Human geographers tell us that places happen at the confluence of the *physical* stuff of the world, our *social* relationships with it, and the *meaning* we ascribe to those relationships. Essentially, places are about discrimination, or selection, if you'd rather. We say: the set of relationships I experience and the values I assign *here* are different from those I experience and assign *there*.

As a result—we might even say by definition—research findings about place rarely generalize from one place to another. Place name itself is enormously valuable in Champagne, but it hardly follows that the same will be true in the Okanagan Valley—which, if you're unaware, is Canada's second-largest wine-producing region, responsible for pleasant reds from British Columbia.

Geographic indications don't all indicate geography in the same way. It should therefore come as no surprise that grape-growing regions *are* occasionally defined on the basis of soil itself—yet even in these cases, soils aren't the whole (or only possible) story. New Zealand's Gimblett Gravels is a striking example. New Zealand only instituted a country-wide geographical indication system in 2017, designating the Gimblett Gravels as a subregion within the appellation of Hawke's Bay. "The gravels," however, have effectively been defined by local edict since shortly after the first vineyards were planted on this then low-value (from a sheep-growing perspective) land in 1981. When a proposed

---

[4] Cross, Plantinga, and Stavins, "What Is the Value of Terroir?"
[5] Keating, "An Empirical Analysis of the Effect of Subdivisions."

quarry threatened to wreck the area's flourishing viticultural prospects in 2001, the Gimblett Gravels Winegrowing Association won over the local council, trademarked their name, and attached it to a specific region drawn up from a map of soil types.

To use Gimblett Gravels on the label, a winery must demonstrate that at least 95 percent of a wine's grapes come from a vineyard whose soils are at least 95 percent composed of soils derived from the Ngaruroro River bed: Omahu, Flaxmere, and Omarunui soil types. That requirement says something about the values of the local community, the distinctiveness of their soils, and the relationship between the two—that is, about the stories this community wants to tell about their place. But do we have evidence to support that these soils are the single most important factor in shaping the region's wines?

Naming Priorities

Identifying which factors most contribute to the characteristics of a particular *terroir* in a scientifically rigorous way is a huge, tangled problem. 🖉 Science generally relies on reducing complex wholes to simpler parts so that the parts can be made sense of and then stuck back together to make sense of the whole. That process requires disentangling each part from other parts—tricky when the whole in question is sensory distinctiveness generated by relationships among an indefinite array of potential contributing factors. An additional difficulty is that, as with economic studies of the influence of place names, we should expect that viticultural studies of how *terroir* works won't necessarily generalize; even the most comprehensive studies can't account for *everything* that might be significant in making one place different from another, such that findings about one place won't necessarily hold true somewhere else.

One of the few generalizable principles to be picked out of that muddle is attached to water availability. The quantity of water available to a vine throughout the year connects to grape size, ripening speed, sugar concentration, color development, and so on. Water availability, in turn, connects to how much precipitation falls, humidity, how far roots must reach to hit the water table, and soil structure. Soil water-holding capacity—the quantity of water that any given soil will contain like a sponge—is thus a major factor in *terroir*. Soil water-holding capacity is related to soil structure. Soil structure is what the Gimblett Gravels is all about.

Let's still remember that soil structure is only one factor of many, even if we're just talking water. Do precipitation, irrigation practices, viticultural techniques, humidity, ambient temperature, or the myriad other factors that might influence water availability vary in and around the Gravels? Those questions become less important, independent of whether and how they affect wine quality (as indeed they almost surely do; the Gravels are warmer than the surrounding Hawke's Bay, for example), because folks who make their living there have decided that their story, the characteristic on which they'll pin their distinctiveness, is *soil*.

So, what if you have the story and not the soil? Central Otago, on New Zealand's South Island, offers a good counterexample. The district accelerated from zero to international recognition in less time than it takes to test the merit of a great Burgundy vintage. Informally, it encompasses six sub-regions. Whether to formalize them in legal terms, and how much to talk about them, are open questions. While some think that the region may not yet have the maturity to know exactly where to draw its lines, others aren't sure what kind of story Central Otago is trying to tell with them.

As you come in from the coast, the first substantial growing area you encounter is Alexandra, a dry, flattish valley with a large town by local standards. Cross the hill and drive past the dam and you're in Cromwell, where vineyards enjoy the mitigating effects of Lake Dunstan. At the north end of the lake, Bendigo vineyards see both the hottest and the most extreme weather. Go east toward Queenstown and you'll find the warmer and drier Bannockburn vineyards perched above the Kawarau River. Through the pass farther toward Queenstown, the Gibbston valley has the highest elevation and the coolest climate. These differences are obvious. Central Otago doesn't tend toward subtle.

None of the conversations happening about subregions question whether there are significant differences. The questions instead are about *what* is significant. Soil differences moving up the terraces from the valley floor? Elevation? Big regional outlines? All viable choices, so the question then becomes: what kinds of similarities are most desirable to highlight? That depends on what wineries want to talk about. Some bottle from estate vineyards. Many blend fruit from different vineyards for balance and complexity—and, no doubt, economy and ease. Most have built identities around concepts other than subregionality.

You can only tell so many stories at once, and subregions are expensive in terms of demands on consumer attention. Geographic indication names are too numerous for most wine lovers to remember; attaching even more names to subregions only exacerbates that problem.[6] While some subregions of Napa, such as Howell Mountain and the Stag's Leap District, garner prices higher than Napa proper, others, such as Coombsville, make little difference or even attract lower prices than the Napa norm.[7] Central Otago wineries have succeeded when someone in Newcastle or Prague recognizes Central Otago. Talking about Bannockburn may be asking for too much.

Is Bannockburn a place? For people who live in Central Otago, unquestionably. For oenophiles who have visited, maybe. For someone musing over the wine list in a restaurant on the other side of the world, it may be no place at all.

---

[6] Atkin et al., "Analyzing the Impact of Conjunctive Labeling."
[7] Keating, "An Empirical Analysis of the Effect of Subdivisions."

2

Vines 🍃

Classifying things—any things—has everything to do with our reasons for classifying them. Carl Linnaeus is widely known for advancing the modern idea of taxonomy. But Linnaeus is only Linnaeus when we classify him as the first taxonomist. The Swedish version of his name was Carl von Linné, and were we to talk about him as an eighteenth-century Swedish guy, or a father, or maybe a darn good fisherman, we might call him that instead. The same principle applies to naming grapevines. Whether two vines belong to the same group has something to do with genetics and everything to do with why we're grouping them.

Genetic Inheritance

Let's start with the genetics. Grapevines are highly heterozygous, which is to say that many of their genes come in multiple versions. Humans are heterozygous for eye color; everyone has several genes from each parent that establish their eye color, and human eye colors vary widely (though within a predictable range) as a function of exactly which version of those genes we inherit. Grapevines are heterozygous for berry color and berry bunch shape and growth habit and all manner of other characteristics. Cultivating grapes from seed is therefore incredibly unreliable. You can't be sure which genes any one seed will inherit, in which combination. And because some genes are recessive, only showing their effects when an individual carries two copies, a predictable fraction of the offspring resulting from a grapevine mating exercise may display a trait that wasn't visible in either parent at all.

What I've just described is *Mendelian* genetics, named for Gregor Mendel, the nineteenth-century monk who was sufficiently obsessed with inheritance and pea plants to observe that peas carry the traits of their parents (including those pesky recessive ones) in predictable

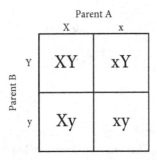

Figure 2.1 A Punnett square, used to determine the ratio of offspring that will end up with a particular set of genetic characteristics from a mating of two parents, *if* the characteristic follows Mendelian genetics.

ratios (see Figure 2.1). Mendel's work was famously ignored for decades before being compiled with newer, fruit-fly-based data about how inherited traits could be traced to the behavior of chromosomes.[1] The result was the "modern synthesis" of Mendelian and molecular genetics, and the foundations of twentieth-century molecular biology.[2]

Since then, geneticists have found that Mendel's delightfully simple inheritance ratios simply don't work for many inherited characteristics. One difficulty is that multiple genes may influence a single observed trait, such as human eye color. Another is that DNA segments sometimes relocate from one chromosome to another, changing the ratio of offspring that end up with one parent's version of a gene versus the other's. Occasionally, too, random mutations lead to something completely unexpected. These and other complications mean that breeding

---

[1] Fruit flies have enormous chromosomes in their salivary glands, roughly a thousand times fatter than the chromosomes found in most cells. The ease of seeing those structures under a straight-forward light microscope made them ideally suited for early experiments in the 1930s that first linked chromosomes to inheritance. Chromosomes include both DNA and proteins that keep that DNA organized, so scientists were unsure about which one was the key genetic material until Watson and Crick more or less settled that question in favor of DNA in 1953.

[2] Historians of biology would have my head for glossing the modern synthesis so superficially, since how this massively important scientific happening happened is a subject of considerable study. For now, I'm trying to get somewhere without getting bogged down in this particular detail.

grapevines for desirable characteristics isn't as predictable as Mendel and his peas might have led us to expect. Mendel happened to identify a few straightforward traits to investigate. The challenge ever since has been dealing with the exceptions.

Mapping genetic relations, exceptional and otherwise, has been completely reformulated over the past several decades by DNA sequencing. The old method required looking for genetic effects in the outward appearance of an organism. The new method lets scientists look at differences in DNA itself. The difference between the two means that DNA sequencing has utterly unsettled the world of taxonomy and raised new complications for establishing exactly where lines between species should be drawn. The idea of "species" itself is a problem for many microorganisms, as Chapter 7 describes. ⬤ Meanwhile, wine grapes come with their own taxonomic peculiarity because virtually every vine comes with two separate sets of genes.

Phylloxera → Grafting

Once a vineyard is established, it's usually fair to think about each vine as an individual plant with leaves above and roots below. But before that, when a vineyard is being planted, that vine comes in two pieces: the rootstock, and an entirely different grape variety grafted on top. The story behind why vines come in two parts is integral to how the contemporary wine world has come to decide which differences among vines are worth naming.

The first part of this tale is well known. In 1866, a tiny, innocuous-looking (see Figure 2.2), massively destructive aphid-like insect traveled from its native home in Missouri to Europe, hitching a ride on transatlantically imported plants and leaving near-total vinous destruction in its wake.[3] Phylloxera, as that insect is now universally known and reviled, attacks a vine's roots. Grapes native to the Americas, having grown up with phylloxera in evolutionary terms,

---

[3] No one is sure exactly how or when phylloxera first arrived in Europe. A vigneron in the Rhône valley first observed phylloxera-affected vines in 1866, but the pest would have had to arrive before then.

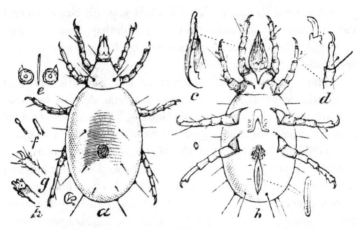

Figure 2.2 A top and bottom view of the phylloxera louse from the June 5, 1874, edition of *Popular Science Monthly*, in which Dr. Charles V. Riley reviewed grapevine pests, of which this critter was undoubtedly the worst. Image in the public domain.

have evolved a technology to ward off their attack in the form of a sticky sap that effectively glues aphids' mouths shut when they try to feed. European grapevines lack such defenses, understandably, as they had no reason to develop them before unwitting humans recently introduced one.

As phylloxera ate its way across Europe, starting with France, long-established vineyards turned to dust. Meanwhile, scientists and officials argued for years about whether the insect was the direct cause of the plague or an indirect symptom of some kind of long-standing vine imbalance, largely because this was a transitional time in medical science. Germ theory—the idea that an outside agent such as an insect or a microbe might be the root cause of disease—was still less popular than blaming illness on some fault in a creature's internal constitution. (Louis Pasteur, who would later become a national hero for convincing France and the rest of the world that germ theory was correct, still had ten years or so before he would conclusively win that fight.) After years of arguing, the germ theorists did eventually win out over the constitutionalists, but even then, how best to thwart the louse remained an issue. Though some authorities advocated for chemical fumigants and others dug holes around vines and filled them with sand (phylloxera doesn't like sand), grafting European vines onto resistant American rootstocks eventually prevailed as the mitigation strategy of choice.

Grafting was necessary because wines made from American grapes don't taste very good, at least not to palates trained on European wines. American grapes differ sufficiently from their European cousins to be considered a different species, but the two are similar enough to mate. The results are often at least partially phylloxera-resistant, so hybridizing could have been a solution for replanting Europe. However, while wines made from American-European hybrids improve on those from their American parents, their flavors remain so far from European varieties that replacing one with the other was hardly a palatable option. In contrast, grafting European varieties onto American or hybrid roots delivered European characteristics and American pest resistance at the same time. Thankfully, initial fears that American rootstocks would pollute the flavor of European vines grafted onto them proved unfounded.

The vast majority of the world's wine-destined vines are now the product of having literally been cut in half and tied back together.[4] In the long term, the worst casualty of the phylloxera crisis may be that we'll never know how many marginal, unsung European varieties were lost because they were never grafted and preserved. Of course, some varieties that gave growers less to sing about may have been left behind on purpose.

I'm rehashing this story because before phylloxera became a crisis, precisely identifying grapes was less important. A European vineyard managed by the same family for generations might contain a whole mess of different vines. Some might not even *have* names. Phylloxera and grafting reshaped the landscape not only of European vineyards, but also of grape classification.

---

[4] Australia, portions of South America, and pockets of vineyards elsewhere remain phylloxera-free and own-rooted.

Varieties → Clones

What we call varieties come from mating two vines and planting the seeds. Pinot Noir, Chardonnay, Mourvedre, Merlot, and every other even modestly known variety exist because seeds *aren't* the typical way to plant a vineyard.[5] Because grapevines are so highly heterozygous, vines grown from seed are highly unpredictable. Consequently, far prior to phylloxera, most new grapevines came about, not from seed-producing sexual reproduction between two vines, but from vegetative asexual reproduction between an established vine and a human who chose to clip a branch and stick it in the ground where it could grow new roots. Vegetative reproduction makes it possible to directly select and propagate adult vines with desirable characteristics. As a result, instead of an innumerable mishmash of varieties cropping up every time a vineyard is planted, a handful of favorites dominate the world over, with only a thousand or so named varieties in active use altogether. Very few people are interested in the unusual game of Russian roulette known as planting vines with unpredictable characteristics from seed.

Post-phylloxera, establishing a vineyard required more specificity, both in one's choice of rootstock—which must be matched to the vineyard's characteristics—and in the variety grafted on top. Indeed, variety isn't specific enough. Now we have to talk about clones.

"Clone" may bring to mind B-list sci-fi or Dolly the sheep, but cloning a plant such as a grapevine relies on a wholly different protocol than cloning an animal. Cloning an animal calls for removing the genome from an egg or other cell similarly capable of differentiating into a new organism, replacing it with the genome of the animal you want to clone, and then taking a lot of care to convince the result to grow. Cloning a plant merely requires that the plant be able to reproduce vegetatively, from a cutting, because a cutting will ordinarily be genetically identical to its parent. That's the general rule. Grape varieties come in *clonal selections* because of exceptions to the general rule.

---

[5] "Variety" is a distinct version of a domesticated plant or animal, though "breed" is more commonly used for animals. "Varietal" is an adjective describing wines made from grape varieties. By convention, the names of grape varieties are capitalized; varietal wines made from them are not.

Clonal selection is warranted by the brutal reality that all creatures suffer genetic mutations. Exposure to UV radiation, chemical insults, and the mundane tendency of DNA replication machinery to make mistakes all mean that individuals' genetic material accumulates myriad tiny changes over its lifetime. Most of those changes don't do anything. Many are fixed by DNA repair equipment, something like a road maintenance crew and a team of expert proofreaders folded together. Many that remain are inconsequential because changes to DNA often don't translate to meaningful changes to the proteins and RNA molecules that get things done around the cell. And for humans and other creatures that reproduce sexually, mutations are only passed on to the next generation when they happen in a sperm or egg cell, which is rare.

Mutations are mostly inconsequential for grapevines and other vegetatively reproducing creatures, too, but they propagate differently. A mutation in any cell that divides and grows into a section of a plant can be reproduced via a cutting from that section. Again, most won't yield any significant change at all, and a majority of those that do will yield unfavorable changes. But, very rarely, a mutation results in a visible improvement: a vine that's bigger, stronger, faster, a different color, or otherwise a source of variation that a forward-looking grower might want to maintain.

Deliberately selected mutations are responsible for the delightful group of grapes that we know and love as Pinot. Pinot Noir, Pinot Gris, Pinot Blanc, Pinot Meunier, and their lesser-known relatives are all one variety in the sense of having an essentially identical genetic background derived from mating the same two parents (though the lineage is sufficiently ancient that those parents can't now be identified). Over that lengthy history, Pinot has developed multiple mutations that affect anthocyanins, the set of compounds most directly responsible for grape color.[6] Pinot Noir berry skins have two layers of anthocyanin-containing cells. In Pinot Gris—"gray" pinot, though the grapes are a pink-y bronze—one of those two layers is missing. In Pinot Blanc or White Pinot, both layers are missing. The precise type of mutation involved in

---

[6] Mutations aside, all grapes have the machinery to be red. The same kinds of mutations that disable that machinery seem to exist in all white grape varieties, though it seems likely that they arose on more than one independent occasion (see Vezzulli et al., "Pinot Blanc and Pinot Gris").

Pinot's case can easily flip-flop as individual cells divide, making multi-colored grape bunches fairly common sightings in pinot vineyards.

Pinot has also accumulated mutations that have been found worthy of propagating for other reasons. Pinot Meunier is named for its hairy leaves; *meunier* means "miller" in French, and the variety's white hairs look like a dusting of flour on the undersides of leaves. It's also fruitier and a bit earlier to ripen than Pinot Noir. Others have been selected for different growth habits (Pinot Droits) or fruiting patterns (Pinot Fin and Pinot Moyen).[7]

University viticulturists intentionally establish some new clones. Others become well known when someone borrows—or smuggles, according to popular legends—them from renowned vineyards. For instance, the formal name of a Pinot Noir clone favored in New Zealand is "Abel," after the customs agent who found it and forwarded it to university viticulturalists, but it's informally known as the "gumboot clone"; as the story goes, it was imported into New Zealand hidden away in a gumboot, the local term for tall rubber footwear plausibly capacious enough to contain your leg and a snippet of grapevine at the same time. Pinot Noir "Dijon" clones are so called in the United States because they were shipped to Oregon from a Dijon address in the 1980s; in Australia, they're better known as "Bernard" clones after the viticulture professor in Dijon who was doing the shipping. Clones with numbers, such as the charmingly named Chardonnay 08, tend to be the product of deliberate academic efforts. In number eight's case, the responsible party is Foundation Plant Services at the University of California, Davis. Even numbered clones, though, originally came from someone taking a cutting of an unusually appealing bit of vine.

These origin stories are fun, but they don't carry much explanatory power. Better explanations are less linear. Because many (albeit fewer, post-phylloxera) esteemed European vineyards remain *selections massales*, comprising a mass of minutely different vines, selecting

---

[7] According to Jancis Robinson, Julia Hardin, and José Vouillamoz's magnum opus *Wine Grapes*, Pinot Droits had a moment in the 1960s when the variant's large berries appealed to vignerons who prioritized quantity over quality, but those vineyards have now been planted over with smaller-berried selections that yield more concentrated wines.

and propagating any one of them is hardly representative of the original mix. On top of that, vines don't yield identical produce everywhere; their characteristics are a function of interactions with their environment. Deriving your vineyard's genetic material from one vine that happened to contribute to a first-growth Bordeaux is likely to be as much about reproducing a story as reproducing that vine's characteristics.

Distinctions among clones and varieties say as much about us humans as they do about the plants to which we attach them: because we can only see what our (microscopy, DNA sequencing, or unaided-eye) technologies enable us to see, because we can only make collective sense of differences when we have language to talk about them, ⸲ and because someone has to decide which differences matter enough to warrant a name. How lines are drawn among one and the next is cultural—by way of their names, to be sure, but also because their underlying "natural" genetics are already a matter of cultural values.

But Which Differences Make a Difference?

In the 1930s, distinguishing one grapevine from another usually came in the form of ampelography, literally "vine-writing," or examining leaf shape, cluster dimensions, and other observable physical characteristics.[8] Botanists and horticulturalists used vine-writing differently. Botanists, whose business is classifying plants, needed to establish family relationships among varieties. Horticulturalists, whose business is caring for plants, needed to create practical guides for distinguishing varieties. As a result, we have taxonomic guides and field guides—two very different things, as anyone who has mistakenly picked up one when they really needed the other can attest. A taxonomic guide tells

---

[8] As Frederic Bioletti introduced the subject in a 1938 article in *Hilgardia: A Journal of Agricultural Science Published by the California Agricultural Experiment Station*, "The species of the botanist is a group of like individuals, usually seedlings; the variety of the ampelographist is a single individual—the clone—or the totality of all the plants derived by vegetative propagation from a single seedling and constituted therefore simply of parts of the same individual."

you who's related to whom. A field guide tells you what to call the cute fern you saw while hiking so that you can point it out by name when you bring a friend next time.

Ampelography has now been firmly replaced by DNA sequencing, at least for formal judgments about genetic similarities and family trees. (Leaf and cluster shapes are still more useful for identifying vines in the field, for obvious reasons.) The inordinate utility of sequencing technology is a settled matter. How to use it isn't. In the first instance, there's the question of *what* to sequence: which organisms, and which sections of DNA from those organisms. As sequencing becomes steadily cheaper and more accessible, sequencing all of an organism's DNA, its complete genome, becomes easier. However, it remains the case that scientists often examine just a few targeted regions of a genome if, say, they want to expedite (and economize on) comparing many individuals. And even when whole genomes are sequenced, scientists need to deliberately choose which elements of the resulting vast stream of data to analyze.

Technological shifts challenge categories without offering easy answers about how to redefine them. Should clones be distinguished strictly on the basis of genetic relationships, or is their origin story relevant? What about their growth habits in the vineyard or the qualities of the wines they yield? DNA doesn't automatically win this game. Defining variants is a function of *why* they're being defined, because prioritizing which differences matter requires that we have a reason for caring about why variation matters in the first place. The wine community can't just pass this particular buck back to geneticists. Where lines should be drawn is about values, about which differences are most important. Values questions can't be answered by science alone.

You might therefore think that we're done with what science can say about naming vines, but we're not. New reasons to care about genetic differences are cropping up in the agricultural experiment fields where grape geneticists are trialing new varieties, crafted with the detailed information that DNA sequencing provides. Researchers hope that "ampelographic scouting"[9] among the world's complement of

---

[9] Pastore et al., "Genetic Characterization of Grapevine Varieties."

grapevines, wild and cultivated, will turn up rare examples of vines with genetic resistance to powdery mildew, Pierce's Disease,[10] and other threats that seem to be intensifying with global travel and climate changes. When they find those examples, new-fangled DNA sequencing can guide old-fashioned breeding to move disease-resistance genes from grapes that don't make great wine into a variety that does. The result might be a variety that's 95 percent Pinot Noir plus 5 percent, genetically, of an American or Asian variety you've never heard of.

Disease pressures are also behind new hybrids—full-on crossbreeds of popular wine-appropriate European varieties with an American or Asian parent to yield a variety that's new, tasty, and disease-resistant all at the same time. Notably, as the continent has tended to be hybrid-averse since phylloxera days, a group of fungus-resistant hybrids known as PIWI varieties are being approved and planted across Europe, especially in Germany and the Czech Republic. Just south of Santa Cruz, California, winemaker-rogue Randall Grahm has taken the unusual—some might say crazed—step of planting his Popelouchum vineyard with something on the order of 10,000 new varieties, bred from a mix of "noble" European and "ignoble" disease-resistant parents, in hopes of finding the Next Big Grape and more complex wines at the same time.

A global consensus about what to call the results of such efforts remains pending.[11] Since phylloxera, variety names on labels have become massively important to how the world shops for wine. More than one fortune may rest on whether wine from vines that are, genetically, 95 percent Pinot Noir can be marketed as "pinot noir." PIWI varieties carry the disadvantage of unfamiliar, consumer-baffling names such as Baron or Cabernet Blanc. Then again, they're also marketed as more sustainable and involving few or even no chemical inputs in the vineyard—a step beyond organic—and as such are developing their own niche followings. Novelty may be a marketing vice or a virtue

---

[10] Pierce's Disease is a lethal bacterial infection, of *Xylella fastidiosa*, spread by a small insect called a sharpshooter. The insect deposits the bacteria in a vine's circulatory system when it feeds, and the bacteria block that circulatory system, often killing infected vines within a few years.

[11] Secretary of the General Assembly, "OIV Process for the Clonal Selection of Vines."

depending on who buys your wines. We're back to values. No rule will treat everyone's interests equally well.

Biodiversity is also at stake in this conversation. Even post-phylloxera Europe is home to a gobsmacking array of local grape varieties known only to local villagers and sleep-deprived Master of Wine (MW) students obliged to memorize them. Robinson, Harding, and Vouillamoz found 1,368 unique varieties making some kind of contribution to global wine production in 2011, though most are mere drops in a bucket of global Cabernet, Chardonnay, and other so-called noble varieties. Most aren't famous for at least one good reason. Even oenophiles like me who are easily bored by familiarity have to acknowledge that major international varieties have become international because wines made from them tend to taste good across a wide range of regions. Nevertheless, preserving everything else has a function beyond bored drinkers' niche obsessions. In addition to the less tangible virtues of biodiversity, geneticists are scouting rural corners of the wine world for genes that might be usefully worked into mainstream varieties.

Clones, as they've been developed through the twentieth century, have been criticized for contributing to the monotony of contemporary wines. Where a *selection massale* may compile diverse flavors in a single bottle, wines from vineyards planted to one clone may seem less complex. Worse, when whole regions are dominated by one clone, as has happened with "Gin Gin" Chardonnay in Western Australia or Wente FPS1 Sauvignon Blanc in California, any peculiarity in that clone's susceptibility to disease or environmental change sets the stage for regional disaster.[12]

New varieties such as the PIWIs provide growers with more diverse options. Yet, if just a handful of new disease-resistant Pinot Noir clones emerge from the current crop of advanced breeding protocols, options for risk-averse vineyards may narrow, not expand. Grape varieties may be at something of a crossroads. Genetics alone can't tell the industry which way to turn.

---

[12] FPS are the initials of the UC Davis Foundation Plant Services, who were initially responsible for this selection.

3

Terroir 🖐

My spell-check wanted to turn *"terroir"* into "terror" before I taught it not to. I might have left well enough alone. No other wine concept seems more likely to star in a horror film, though the role it should play is ambiguous.[1] Is it the shadowy *thing* our hero can never quite properly see, because everyone knows it's there but can't quite figure out what it is? Is it the waking nightmare caught halfway between dream and reality, because research wavers between results that say it's real and results that say it's not? Or is *terroir* the hero as they rise above everyone who doubted just how tenacious they could be, because after all of the drama, the concept is as strong as ever? Maybe the best way to play this analogy is to observe just how compelling that sense of mystery can be, and how much it turns off some people who prefer certainty to suspense.

Is *Terroir* a Thing? (What Kind of a Thing Is *Terroir*?)

Terroir is a way to talk about how the characteristics of a wine reflect the characteristics of its origins. Ostensibly, Anglophones don't translate *terroir* because it has no English translation. The penchant of some, primarily English-speakers, to try to sound fancy by using French words might also be a factor. Even so, rough translations point to a bigger problem. "Soil" is the most literal. "Sense of place" is better. Emile Peynaud, an oenologist who brought science and systematization to twentieth-century French winemaking, called *terroir* "Mother Nature." In 2005, France's Institut National de la Recherche Agronomique suggested that *terroir* also involves human culture,

---

[1] Ironically, I learned shortly after writing this paragraph that *terroir* does indeed star in a horror novel of sorts, though I can't tell you which one without spoiling a plot point.

including cultural conventions about which grapes to plant where. In other circles, *terroir* can be a polite euphemism for "this wine tastes like dirt," or a shorter way to say: "I know this wine is fancy but I don't know much else." Most practically, it's a way to discuss the connections between a wine's sensory properties and its provenance, or—and the difference here is significant—how wines from one place are perceptibly distinct compared to wines from somewhere else. The first is a matter of the stories we tell (which doesn't make it any less real). 🐚 The second is something that science should be able to document.

In this conversation, "soil" is a metonym, a rhetorical strategy of citing one element of a thing as a stand-in for the whole. "Lend me your ears" is a metonym; anyone saying it wants your attention, not the cartilage holding up your spectacles. Articulate acts of politeness aside, to say "This wine has great *terroir*" isn't to say that it tastes like dirt, but that the experience of tasting calls up the place where it was produced, its physical characteristics and the expertise and care of its producers. Soil stands in for the uncountable factors associated with a vineyard that would be inconvenient to continually list.

I just called the factors contributing to *terroir* uncountable, but whether or not they are countable—even whether they should be counted—is a research question. Famously, some wine scientists have claimed that *terroir* doesn't exist at all because they *haven't* been able to quantify it. Certainly, as an expression of connection, the concept hardly fits typical scientific paradigms. It's a gesture, a feeling, an attachment, maybe an aesthetic, but not an analytic, not a tightly defined tool, not (it has seemed) a measurable quantity. This last point is at the heart of the issue. Research aimed at documenting *terroir* in quantifiable terms keeps coming close without hitting the nail on its head.

In a robust example, agricultural scientists recently tried to establish whether proposed subregions in Western Australia's Margaret River valley are quantifiably distinct—or, as they put it, to help wine producers gain "a better understanding of the terroir of their region" by determining "whether the variation in soils and climate suggests that subdivision of the region may have merit."[2] That's a massive question

---

[2] Bramley and Gardiner, "Underpinning Terroir."

from a scientific perspective, considering what measuring "soil" and "climate" entails. They could get a handle on the first one thanks to the Soil and Landscape Grid of Australia, a (nearly) whole-continent map of soil and land characteristics at football-field resolution, from the surface to a depth of about two meters. This is an avalanche of data, even if one set of measurements per football field is low-resolution from a vineyard perspective. On top of all of that, to address the "climate" half of the question, the researchers drew on thirty solid years of regional weather reports.

From that mountain of measurements, the researchers chose ten metrics covering temperature, temperature variation during the growing season, soil drainage, soil nutrient-holding capacity, and elevation. No matter how they sliced that dataset, they couldn't resolve subregions with distinct physical characteristics. Soil characteristics weren't helpful because soil type varies in small pockets across the breadth of the valley.[3] Temperature was better (though still not good enough), maybe because higher temperature correlated with lower rainfall, so one measurement effectively accounted for two. But no matter what they tried, they couldn't identify a soil- and climate-grounded way to explain how the characteristics of one region cleanly distinguished it from the next. In the end, they concluded that they probably needed a finer-grained view of the land, given all of those soil differences. They also concluded that they probably needed to account for variations in local vineyard management practices, and for how the wines taste.

Other research teams have begun with wine quality but haven't ended up with clearer results. When food chemists searched for molecular fingerprints that might distinguish grapes and wines from subregions of Rioja, they just barely found patterns that linked *grapes* with regions.[4] They couldn't find the same patterns for finished *wines*—which are, of course, what tasters taste when they say that subregions have signature flavors. In another example, a Bordeaux-based study tried to match Bordeaux subregional grapes to subregional soils. This

---

[3] Wine-growing places frequently host a mix of soils in close proximity, but some distinctive regions are indeed characterized by a distinctive soil type, such as the Gimblett Gravels on New Zealand's North Island.

[4] López-Rituerto et al., "Investigations of La Rioja."

effort found alignments between soil type and grape sugar content, but not grape pH or acids.[5] Meanwhile, a Burgundy-based study found that vintage-to-vintage variation was a much more powerful influence on wine's molecular composition than its provenance.[6]

Before we make too much of that lack of conclusive evidence, we need a better idea of what we're looking for. Research is bound to turn up little if research methods are out of whack with what you're trying to find. An archaeological dig for unicorns may be set up to fail, but you'll learn a lot about the species if you go digging in thirteenth-century French literature. It's easier—and, dare I say, more scientific—to believe that research hasn't been asking quite the right questions than that a very large number of winemakers and wine drinkers are enthralled in some kind of mass delusion. Admitting that former possibility, we need to do some more work to make sense of what kind of thing *terroir* might be.

Most contemporary scientific investigations into *terroir* begin from the same starting point as contemporary scientific investigations into anything: with the idea that the object they're seeking to outline is decomposable, influenced by key indicators, and a function of physical natural phenomena. Each sets up assumptions about the kind of thing *terroir* is presumed to be.

Decomposability. The reductionist mode that dominates most scientific fields operates under the assumption that complex things are best studied by breaking them into parts and studying the parts individually. Their function after disassembly may not be identical to their function in the whole, but the assumption is that characterizing them separately will nevertheless yield meaningful knowledge about how they work together. This approach works beautifully for things that were assembled from parts in the first place, like sandstone or toasters. It works less well for yeast cells, grapevines, and many other things not made by humans. When scientists decompose those things, they lose something in the process. They also must *decide* what constitutes a part. And no matter what decisions they make, they'll never isolate pieces that preceded the whole (see the MACHINE LEARNING box).

[5] Leeuwen et al., "Influence of Climate."
[6] Roullier-Gall et al., "How Subtle Is the 'Terroir' Effect?"

MACHINE LEARNING

Machine learning is beginning to mitigate some of the limitations of parts-based science. Neural nets can try out umpteen different ways of arranging a massive (or gargantuan, or even merely big) set of data, finding patterns in the whole that a human analyst would never have been able to find, let alone on a manageable timescale. Artificial intelligence can also try out innumerable ways of putting parts back together *in silico*, spitting out strategies for assembling an award-winning powerhouse cabernet sauvignon just as well as talking heads in deep fake videos. What machine learning and AI can't do is remove either the human or the part from the equation. Someone has to write the algorithms that tell the machine what it's trying to learn, shaping what the machine spits out. Outputs of machine learning are predicated on inputs of machine learning, and the data we have to feed to the machine generally come from studies of parts. And at the end of the day, someone has to decide what the outputs mean and how to use them.

Decomposability is an overarching framework for biology for a reason. It's sometimes the only way to experiment with complex living systems, changing one variable at a time to pin down the effects of that variable specifically. Without first decomposing the system, trying to make just one change while holding everything else steady tends to inadvertently entail a whole slew of changes, such that matching a specific cause to a specific effect is impossible. Changing irrigation regimes doesn't just change how much water a vine receives, for example, but also soil temperature and nutrient availability and cover crop growth and any number of other relevant concerns. Disconnecting those connected factors makes it possible to link grape color development or sugar accumulation, say, to water availability or soil temperature specifically, but simultaneously distances the results from how vineyards ordinarily function as complex wholes. Applying decomposability to *terroir* therefore embeds the assumption that *terroir* can be understood through countable parts that, if not initially discrete, can be meaningfully made that way.

Key indicators. Most natural-scientific phenomena are complex, with many moving parts. Measuring many moving parts is resource-intensive. Consequently, researchers often look for key indicators: one or a few measurements that disproportionately inform the state of the whole. Blood cholesterol is a key indicator for a person's risk of developing heart disease. The causal connection between high cholesterol and heart disease is iffy. Particular forms of cholesterol known as low-density lipoprotein (LDL, the "bad" cholesterol) can accumulate in blood vessels, stiffening and narrowing them, making the heart work harder and leading directly to disease. ◢ However, cholesterol won't become sticky in the absence of platelets activated by inflammation. Recent evidence has suggested that inflammation itself may be the most important element of that equation. However, high cholesterol statistically correlates with heart disease risk, and blood cholesterol levels are measurable warning signs. Some key indicators *are* causal at the level of individuals, but in practical terms, what matters most is that they're reliable at the level of populations.

Terroir research often seeks a measurable parameter or two—rainfall, soil composition, or soil temperature, for example—that might reproducibly associate with the difference between one wine-growing region and the next, without researchers needing to sum up dozens of separate measurements to explain that difference. Dozens of factors may contribute to regional distinctiveness, but if two of them are sufficient to draw lines on the map, making two measurements is more efficient than making twenty and maybe nearly as effective. Applying the idea of key indicators to *terroir* says that this phenomenon may have a lot of parts, but some parts are more important than others, so measuring just those parts should reveal the same patterns that we'd see if we'd generated a composite metric to describe the whole system. As we can see in the Margaret River example, however, mismatches in resolution can be an issue.

Physical natural phenomena. Here, we're talking about the scope of what scientific research takes into account. If *terroir* is assumed to be a manifestation of a physical location, then it's reasonable to study *terroir* by gathering data about soil composition and soil structure and weather patterns and other physical dimensions of that location. If *terroir* is a function of a *place*, with characteristics that are about cultural practices as well as physical parameters—or, even better, about interactions among cultural practices and physical parameters—then *terroir* research might need to account for what humans do as well as soil and weather and such. We should expect cultural practice to result in physical changes, but if you're trying to get at culture, you might collect data to explicitly account for habit and tradition. Many scientific studies, however—like the Margaret River survey—interpret *terroir* strictly in terms of landscape characteristics. Of course, scientists may well believe that culture is important, but may not well-equipped to account for it. Employing a simplified construct for research purposes doesn't restrict someone to believing that construct to be the whole truth of the matter—especially when decomposability and key indicators suggest that studying just the physical components on their own *may* be enough to explain regional differences—but cultural elements may still end up left out at the end of the day.

These approaches all admit the possibility that *terroir* might be an emergent property that can't be entirely addressed by natural sciences alone. All the same, a study's underlying assumptions are easy to forget when you're busy doing the studying, and sometimes simplifying moves made for the sake of experimental tractability get confused with phenomena themselves.

Terroir could be an emergent property that only manifests when a large number of parts come together. It could be linked to interactions that don't tidily separate into parts. It could be a product of social, cultural, and historical considerations that aren't easily quantified in scientific terms. Decomposability, key indicators, and physical parameters may or may not be enough to get at all of that. But in the end, to make sense of how we *should* be looking for *terroir*, we need to ask a bigger question: what is *terroir* good for?

It's difficult and not very helpful to insist that a particular definition of *terroir* is right or wrong. It's still difficult, but maybe far more helpful, to ask what kind of work *terroir* is good for doing. For my own part, in a fragmented world of fungible commodities, agile labor, and decoupled resources, I want to hold on to *terroir* as a tool to connect lives and experiences and ways of being. So here are my propositions for what kind of thing we might be looking for. Hold on: we're about to do some theory, but being conscious of one's assumptions while doing science means digging into philosophy.

One: *Terroir* is a process, not a thing.

Two: *Terroir* is material and discursive, a function of physical things and a metaphor.

Three: *Terroir* is an emergent property.

Fourth and finally: None of that means that *terroir* can't be studied scientifically.

One: *Terroir* Is a Process, Not a Thing

We're endlessly dealing with things. The keyboard I'm typing on is a thing. The coffee I just finished drinking is a thing. So is the book you're reading and the text document in which I'm currently writing it. We might even agree that both you and I are things.

The trouble with things is that they change over time. That coffee is well on its way to becoming me and my waste. You and I are different people than we were yesterday. My text document is growing as I type, the keyboard will break down eventually, and if this book doesn't decompose because it gets wet or is eaten by insects, it will still eventually become unreadable because language changes over time.[7]

Since the nineteenth century, some scholars have suggested that things aren't things at all; they're processes. A school of philosophy of biology holds that the basic unit of life is the process, because the fundamental reality of all living things is that they're always changing. Some scientists have picked up on a similar idea—that instead of subdividing biology by creature type, biologists should reorganize around how a common grammar of processes unites all living things. The cell membranes that keep life organized might look like nouns, but we can learn more about how they work when we think about them as verbs, constantly shuttling molecules back and forth, opening

---

[7] Goodness knows that if you're reading an ebook, the data can be erased or the server may fail.

and closing ports, communicating with and responding to the outside world.

Verbing is hardly the sole privilege of living things. Landscapes constantly become themselves over geologic time. As I write this book, I exert force on my keyboard, wearing it down in ways peculiar to my typing habits; simultaneously, it enables me to become someone who's writing a book. Even things like ideas can be understood as processes that become in relation to other things over time, because people who think about ideas change the ideas and the ideas change them back.[8]

Thinking about *terroir* as a process, rather than a thing, makes sense of how it describes connections. The elements of a place, (conventionally understood as) living and otherwise, are always in the process of shaping that place and one another in an endless set of concatenated loops.[9] Viticultural practices influence soil temperature influences microbial life influences soil structure influences water availability influences root structures influence viticultural practices over time. Making sense of *terroir* can't rely on the idea that places *have* characteristics. Instead, places develop characteristics as they continue to become.

Two: *Terroir* Is Both Material and Discursive

Whether you think of things in nouns or verbs, you're likely to think of things in terms of material stuff. We can also describe every thing in terms of meaning, or the kinds of relationships they form that build structure into the world—that is, their *discursive* qualities. That's to say, everything can be described in physical dimensions *and* in discursive

---

[8] Centering processes doesn't mean giving up nouns. That would be impossible, given the structure of English and other widely spoken languages (though less so in some less widely spoken ones, such as Navaho). But because the structure of our lingua franca tends to encourage talking about moving through the world as though we're talking about interacting with stuff-out-there with fixed identities, would-be process-talkers have to work a bit harder to mold language to our purposes.

[9] *Becoming-with* is a conceptual research tool attributed to Donna Haraway, an interdisciplinary researcher of creaturely relationships. If you're interested in how humans and other creatures get along—or don't, in these troubled times—Haraway's exceptionally popular and unusually accessible book *Staying with the Trouble* is a thought-provoking read.

or meaning-making dimensions. We can also describe all things in terms of energy, as trajectories of movement through time as they continue to become themselves in relation to others. Cell membranes are physically composed primarily of phospholipids, ⚛. but are meaningfully the leaky boundary between cell insides and outsides. Energetically, they're a bit like a battery, creating an electrochemical difference between two spaces that drives molecular movement from one to the other.

Physical, discursive, and energetic dimensions of things are all connected. How we make sense of things influences what we do with them and how we interpret data we collect about them. Physical features of the world continually shape how we talk and make meaning, in endless iterative loops. Thinking of cell membranes as *barriers*, for example, has long emphasized their stable wall-like properties over their dynamic communication properties.

Terroir is a way to make sense of observable phenomena, and a way to tell stories about how wine meaningfully connects people and places and experiences. It's a physical phenomenon grounded in measurable material things. It's an idea about how to understand the world that exists in language. It's a process that moves through time with direction and momentum. These aren't mutually exclusive.

Three: *Terroir* Is an Emergent Property

"Emergent property" is shorthand for saying that some things (or processes) are more than the sum of their parts, and that parts don't necessarily explain the behavior of wholes. If we decompose humans into individuals, we lose societies. If we decompose brains into neurons, we lose memory and identity. In systems with emergent properties, decomposability doesn't tell the whole story.

Studying place-based effects as individual properties is difficult precisely because they're so densely interconnected. *Terroir* research rarely involves experimenting with parts in isolation by, for example, planting a specially designed research vineyard with identical vines rooted side-by-side in distinct soil types. That approach is expensive and technically challenging; plots of land have an awkward habit of incorporating uncontrollable variation. But it also mightn't be very helpful if the flavor of a place relates to how soil structure *interacts* with other characteristics. Isolating potential causal factors might eliminate the very thing you're trying to cause.

Take the soil surface. The superficial appearance of a vineyard's soils—what the ground looks like as you're walking on it—can be a highlight of *terroir* stories, especially in vineyards covered in chunky rocks that look wholly improbable for growing anything. Even if it's only skin deep, a soil's surface appearance matters because whether it's light or dark, shiny or dull changes how warm it gets. The temperature of the soil surface, and its reflectivity—how much sunlight it bounces back up into the canopy of leaves above—matter to leaf and grape temperature, which matter to ripening and to how ripe grapes accumulate flavors and colors.

So what matters to the soil surface? Not just the characteristics of the soil as such, but how it's been managed. Have cover crops been planted or native vegetation allowed to grow, or is the vineyard tilled or (and) sprayed with herbicides? Has anyone laid mulch? Soil is shaped by how long the vineyard has been a vineyard and what it was before, and whether tractors or herds of sheep are run through it. And returning to iterative loops, how the vineyard is managed is at least in part a function of the types of soils it contains.

This knot can't easily be disentangled into something that runs in a straight line. Studying soil surface alone, while trying to control these other variables, might not tell you much. Studying soil surface in its interconnections changes lots of things at the same time, so deciphering precisely what matters is difficult.

Four: *Terroir* Can (Still) Be Studied Scientifically

The deeper science digs, the more it's clear that not everything can be understood by picking it apart. Dissecting a cell to characterize each molecule has yet to comprehensively explain how life works because cells are no more static than places are. They endlessly become, changing from one moment to the next as their parts bump against each other and their environments. The solution isn't to throw out analysis. Relationships can be subjected to analysis, too, but doing so requires thinking about what can be measured to get at interactions. Among many other things, for *terroir*, that means microbes.

If you speak microbe, you'll find communities of creatures in cacophonous conversation in most places, and soil especially. When a wasp or a tractor moves from one place to another, it carries microbes with it. When vineyard managers choose an irrigation regime, or decide whether to spray herbicide or grow a cover crop, they alter microbial communities. Were you and I to shake hands, we would exchange microbial cells dwelling on our palms in addition to greetings. The microbiome on my hands is a physical record of the people and things with which I interact. In short, microbes are in the midst of how creatures and soils and animals and plants and other things interact, and how they change each other by interacting.

Microbial *terroir* has taken off as scientists increasingly appreciate just how central microbial life is to everything else. More than just appreciation has been at work; scientists have also been wielding new tools, like cheap and easy DNA sequencing, that make it possible to pick apart microbial communities that remained invisible until just a few decades ago. 🐝 In wine, we can date this microbial turn to 2013, when the publication of a study by a team at the University of California, Davis used then-new DNA sequencing strategies to illuminate a previously unseen view of wine-region distinctiveness.

A graduate student named Nicholas Bokulich and his supervisor, Professor David Mills, cataloged yeast and bacteria from vineyards across Napa, Sonoma, and the Central Coast of California, screening freshly crushed grape samples for bits of DNA that signal the presence of particular microbial species.[10] The moment at which they sampled was significant. By mapping microbial populations immediately after grapes are crushed and before fermentation begins, they accounted for microbes hanging out in the winery or associated with the equipment (and staff) that brought those grapes in, in addition to those from vineyards themselves. In other words, they accounted not just for

---

[10] Bokulich et al., "Microbial Biogeography of Wine Grapes."

physical landscape characteristics *or* for cultural practices, but for how these things come together.

The microbial communities they uncovered grouped remarkably well into regionally distinctive patterns; Napa, Sonoma, and the Central Coast all had distinct microbial signatures. In 2016, Bokulich and Mills were even more specific, showing that within the Oakville region of Napa, the neighboring wineries of Far Niente and Nickel and Nickel hosted similar but distinct microbial communities. Importantly, each winery's cabernet sauvignon and chardonnay—both made as they typically would be for sale—held on to those unique microbial signatures from crush through to the chemical signatures of the finished wines.[11]

By 2016, microbial *terroir* had become fashionable, with a steady stream of publications documenting the unique microbial profile of the world's wine-producing sites. But this kind of research is tricky, and not just because of the *terroir* part. Knowing where a microbe *is* doesn't necessarily tell you what it's *doing* there. Compared to animals and plants, microbes have exceptionally fluid identities. Humans and other animals are born with a particular set of genes and are pretty well stuck with them throughout their whole lives. In contrast, bacteria and yeast often exchange fragments of DNA and, with those fragments, capabilities.[12] As a result, using DNA sequencing to confirm the presence of a microbial species doesn't give you a comprehensive resumé of its skill set—skills that may have to do with grapevine health, or with the flavors imbued to fermenting grape juice. Furthermore, some microbes are better-studied than others. Microbiologists can sequence far more DNA than they can parse in what-does-this-do-in-the-real-world terms.

And yet even if what they're doing remains a mystery, wine microbiologists can still correlate the microbial profiles of a place like Napa, Oakville, or Far Niente with the sensory profiles of its wines. Even if individual microbes can't always be pinned to individual sensory contributions, scientists can still begin to make sense of how microbial communities connect to how wine tastes.

---

[11] Bokulich et al., "Associations Among Wine Grape Microbiome."
[12] This propensity for horizontal gene transfer is part of why antibiotic resistance is so scary, because a resistance gene that evolves in one population of cells can readily spread to others.

At the end of the day, microbial differences should align with differences in flavor, not just with names on a map. Unfortunately, matching crowded, many-specied microbial communities with complex flavor profiles runs us headlong back into the Too Much Data problem. Resolving that problem entails either a lot of computing power (and plenty of research funding) or a return to key indicators, and probably both. Dr. Matthew Goddard's group at the University of Auckland has found success with the simplifying move of assuming that *Saccharomyces cerevisiae* is likely to be the key microbe that matters most to regional flavor. That decision has enabled identifying hundreds of genetically distinct yeast strains associated with individual vineyards around New Zealand that are also associated with their own unique flavor profiles.[13] We have every reason to believe that bacteria and other yeast species also contribute to microbial *terroir*, but what Goddard's group lost in nuance, they gained in feasibility and the ability to see meaningful patterns in the first place.

These studies, and others that are likewise out to account for the physical connective tissue that holds *terroir* together, are adding to what has been an overly simplistic picture of what matters to the taste of a place—a picture grounded in the idea of place as fixed physical land (and maybe climate) characteristics and not much else. We know that picture is overly simplistic because it has proved largely insufficient to explain differences that well-trained humans can taste. What we don't yet know is whether analytical research, as it pulls apart *place* into a larger and more detailed array of parts, will be sufficient to explain *le goût de terroir*. I look forward to finding out.

---

[13] Knight et al., "Regional Microbial Signatures."

4

Minerality 🐚

This chapter is about minerality. It's also about the word "actually," and why you won't find that word anywhere else in this book.

We might be talking about flint or oyster shells, wet rocks or chalkiness, sea air or a zingy feeling, wines that are austere or lean.[1] Expert tasters might use any of these as synonyms for minerality, or consider them subcategories under minerality's umbrella. If those tasters were trained in France, they're likely to perceive minerality primarily with their noses; if they're from New Zealand, they'll more likely describe minerality as happening in the mouth.[2] Some will deride the term as imprecise; others see invoking an impression as saying more about a wine than listing specific descriptors. Casual drinkers in Switzerland are more likely to describe minerality as a salty or zingy sensation compared with their French counterparts, who are more likely to lack words for minerality altogether.[3] You know it if you enjoy chardonnay from the northernmost district of Burgundy known as Chablis, or if you seek out savory, structured whites over fruity, fleshy ones.

Metaphor

Minerality is a metaphor, an effort to make sense of a hard-to-describe sensation by way of comparing that sensation to something else. Most wine descriptors are metaphors. No one is adding cherries or "dark fruits" to their fruity shiraz and cabernet, let alone the tobacco, adhesive bandages, or wet dogs that can appear in tasting notes.[4] The source

---

[1] Parr et al., "Minerality in Wine."

[2] Parr et al., "Perceived Minerality."

[3] Le Fur and Gautier, "De la minéralité."

[4] Dogs, wet or dry, contribute to making many wines, but they're also one member of the production team that never ends up inside a tank; their job description usually involves emotional support with a side of animated doorbell. If your wine smells like wet dog, you're dealing with 2,4,6-trichloroanisole or cork taint, not a friendly multispecies partnership gone wrong.

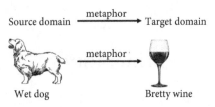

Figure 4.1 Basic structure of a metaphor. "Bretty" wines, infected with *Brettanomyces bruxellensis*, sometimes taste like wet dogs. 🐾

domain for a metaphor (e.g., rocks or cherries) is often reasonably concrete, while the target domain (e.g., the sensory experience of drinking Chablis or fruity Shiraz) is harder to grasp (see Figure 4.1). The first trouble with minerality is that the source domain is hardly obvious. What kind of minerals are we talking about, anyway? Can you pull up a sensory memory of striking a flint? How about tasting one? The second trouble is that unlike those other descriptors I've just listed, sensory scientists and wine chemists have yet to establish a satisfying mechanism that explains why wine might remind you of one.

Mechanism is a relatively recent scientific obsession. In seventeenth-century Europe, when medical care looked like leeches and a lot of impromptu dying, walnuts were thought to be good for the brain because a shelled walnut resembles a shelled brain. We still call a pleasant class of shade-loving plants *Pulmonaria* or lungwort because their spotted leaves resemble the pattern a diseased lung reveals when sliced lengthwise. Past physicians took that pattern to indicate that lungwort was good for treating lung disorders. Pre-modern medicine was fond of analogical reasoning, predicating arguments about the uses of things on what they looked like. The divine creator, so the logic went, had written the name and purpose of every thing plainly in its shape and way of being, such that the observant person could and should learn to read the cosmically ordained meanings of things in this book of life.

Analogical reasoning remains central to modern science. Scientists speculate about newly observed, as-yet-unknown things—lung disorders and rock classifications and the rest—on the basis of their resemblance to previous observations. The key then-versus-now difference is that it's no longer in vogue to assume that gross morphology indicates function, or how something works inside, or its divinely inscribed role in the grand machinations of the universe.

The funny thing is that contemporary nutrition science, with all of its sophisticated molecular biochemistry and data-intensive human physiology, *also* thinks that walnuts are good for brain health—though, today, the rationale is that they're rich in omega-3 fatty acids, which seem to support memory and combat depression in clinical trials. Lungwort remains the subject of investigations into why it so often contributes to traditional remedies for respiratory disorders, because there's some evidence that it works—even though no one has an adequate, mechanistic, molecular explanation for *why*.

Minerality is like walnuts, or lungwort. Some wines remind us of rocks. Contemporary science doesn't have an adequate, mechanistic, molecular explanation for *why*. But that's not to say that minerality isn't real, or that it isn't a useful wine descriptor, or that you aren't tasting it when you enjoy a good glass of Chablis.

Elemental and Geological

To make sense of why, we first need to disambiguate "mineral," a word with two distinct meanings. If you eat a salad as a contribution toward your daily vitamins and minerals, you're dealing in *elemental* minerals such as manganese, zinc, potassium, or sodium. If, after finishing your lunch, you attend a rock and mineral show, you'll appreciate the beauty of *geological* minerals such as feldspar, chalk, or limestone. Wine *contains* elemental minerals; like other living things, grapevines require them to sustain life. Grapevines *grow* in soils that contain geological minerals. We need to make that distinction because minerality is an analogy to geological minerals, not elemental minerals. And yet wine contains elemental minerals but not geological ones.

Minerally wines aren't made by fermenting grapes with rocks; that's certain. Beyond that, we can make three more firm statements. One: minerality doesn't involve tasting components of rocks that vines pick up from soil and deposit in grapes. Two: minerality isn't about tasting elemental minerals. Three: minerality is about tasting something.

Let's start with number one. Chunky geological minerals fragment into fine soil particles through wind, water, the insistent pressure of a growing root, and other forces responsible for erosion. The resulting particles aren't small enough for roots to absorb. Those particles do, however, have far more surface area than their parent rocks. (Elemental) mineral molecules on those surfaces sometimes switch places with molecules in the surrounding groundwater when the result is more energetically stable, leaching (elemental) minerals into groundwater. Plants then absorb waterborne (elemental) minerals when they absorb water. Some rocks and minerals are more likely to dissolve in this way than others—granite less than clay, for example—so the (elemental) mineral content of a vineyard, from a plant's perspective, varies with its (geological) mineral composition.[5]

On to the second point. We can be reasonably confident that minerality doesn't involve tasting minerals because plenty of studies have asked trained tasters to taste minerals, and minerals taste nothing like minerality. If you enjoy drinking mineral water, you know that a high dissolved (elemental) mineral content makes water seem softer or slippery. Mineral water also sometimes tastes salty. This makes sense because salts are, chemically speaking, the coming together of two mineral ions, of which the conjunction of NA^+ and Cl^-, sodium chloride or table salt, is only the best-known example. However, tasting experiments evince that all of these sensations associated with elemental minerals are only discernible, whether in water or in wine, when the liquid in question contains a far heavier dose of minerals than any properly made wine ever does.[6]

Typical mineral concentrations in wine *might* modify overall sensory perceptions in subtle ways that *might* contribute to the sensation we call minerality, but making those jumps lands us in the middle of another problem. The molecular composition of the mineral-laden water that vines absorb doesn't predict the molecular composition of the grapes they produce. Like ours, a vine's circulatory system selectively distributes nutrients to the various parts of its anatomy where

---

[5] Maltman, "Minerality in Wine."

[6] Sky-high mineral content is occasionally reported in wines that have been polluted by corroding pipes or intense automobile fumes from nearby roads. Those situations, thankfully, are rare.

they're needed. Moreover, the mineral content of a batch of grapes doesn't translate directly into the mineral content of wine made from those grapes. Clarification steps remove minerals.[7] Yeast, the bodies of which are removed between fermentation and bottling, ingest some and spit out others.[8]

And here we arrive at the third point. Even though minerality can't be explained by tracing minerals from soil to grape to glass, sensory scientists are reasonably sure that obnoxious wine professionals aren't making minerality up in the interest of making the rest of us feel like ignoramuses. While the associations aren't simple or easy to trace, several studies have identified chemical compounds that consistently show up when experts perceive minerality. Prominent among them are sulfur-containing compounds sometimes also associated with the smell of singed matches or ocean air, which makes a satisfying kind of sense.[9] One of these, methanethiol, also happens to be a signature sensory compound of oysters, the classic, sympathetic pairing for minerally Chablis.

Minerally Mechanisms?

Several studies suggest that sulfur-containing compounds also explain why minerality exploded across tasting notes in the early 2010s. Minerality's rise roughly coincided with a global rise in the popularity of aluminum screw caps. ▌Screw caps block outside oxygen from entering a sealed bottle far more completely than corks do. Wines closed with screw caps are therefore more likely to develop a character typically called "reduced"—that is, the chemical opposite of "oxidized"—associated with a buildup of sulfurous compounds that a trickle of

---

[7] Even though clarification is about removing things, this step also sometimes adds (elemental) minerals when winemakers use bentonite, a type of clay, as a clarifying agent.

[8] Shimizu et al., "Variation in the Mineral Content."

[9] Starkenmann et al. initially set out to study toilet odors, from toilets, when they encountered wine-like flinty aromas by accident. Their main takeaway from the resulting investigation was that minerally sulfur compounds might be unusually difficult to study because they decompose or react while chemists are trying to measure them—a likely-sounding reason why researchers have struggled to pin them down. (Starkenman et al., "Identification of Hydrogen Disulfides.")

oxygen otherwise dissipates. 🔊 Minerality and the cabbage-y, egg-y quality known as "reduced" don't always travel together, but they sometimes do.

Identifying molecules behind minerality is made more difficult by their probably being multiple. Experimental methods excel at associating an experiential phenomenon with a physical factor responsible for that phenomenon. They're less good at associating a phenomenon with a complex network of interacting factors. 🖐 Human molecular genetics is a vivid example. For a time, scientists seemed to be discovering a new "gene for" every week: a gene for blood type, a gene for height, and so on. Of course they were; characteristics and conditions that map to single genes are low-hanging fruit. They're also unusual. Much as intrepid genome explorers might crave a gene for obesity or depression, the vast majority of complex conditions come about through many interacting genes as their products interact with contextual factors that often aren't encoded by genes at all.

Minerality is a complex condition. A few flavors, like vanilla or citrus zest, can be convincingly conveyed by just one molecule (vanillin or D-limonene, respectively), even if vanilla beans and citrus peels contain many additional sensory-active components that add nuance to the naturally occurring versions. Many, like coffee and chocolate—and minerality, it seems—require a mixture of molecules to communicate the basic idea.

If a minerality molecule existed, an intrepid molecular wine explorer would have found it by now. Instead, we have to deal in looser associations. Experienced tasters are more likely to find minerality in wines with relatively more sulfur dioxide and acidity. They're less likely to find it in wines characterized by fruits, sweet flowers, or butter. Minerally wines also tend to be high in malic acid—though minerally wines are for stylistic reasons often not subjected to the malolactic fermentation that converts malic to lactic acid in rounder, buttery styles, so this association may be a tautology. 🦪

Taking the opposite tack, some New Zealand sensory scientists have posited that minerality is about what's *not* there, or that tasters might use minerality to indicate an *absence* of flavor. In other words, they imagined that people pulled out this descriptor when they had nothing else to say. This suggestion arose in the context of observing that French sauvignon blancs were widely described as more minerally than New Zealand ones, and may reveal something about New Zealand wine folk's feelings about French whites as much as anything. In the end, though, when French and New Zealand experts were asked to rate varietal wines from both countries on a line from "absent" to "very strong" flavor and low to high minerality, those characteristics didn't line up after all.[10]

Polarity between fruitiness and minerality might be a function of a different kind of absence: what's *not* in the soil. Minerally wines often stem from austere, rocky vineyards comparatively poor in organic material. Explaining that correlation isn't so simple as saying that rocky soils starve vines while rich ones provide ample nutrients to fuel fruity flavors; organic material isn't universally richer in vine-feeding nutrients, and the extent to which soil nutrients are available for vines to use is another matter entirely. Nevertheless, something appears to be there—or, as it may be, not there—even if what's *actually* going on remains fuzzy.

"Actually"

The word "actually" gets thrown around a lot when conversations turn to analogy, in phrases like "Well, you're not *actually* tasting minerals." What does that mean? Your personal sensorineural apparatus perceives minerals, or minerality. It isn't making that up, nor are you when you put words to that sensation. "Actually" is a way to say that molecular mechanism takes priority over sensory experience as a means of making knowledge about wine. That's a value judgment that could be made differently.

Molecular thinking doesn't align with how senses work. We don't smell and taste in chemical terms. You'll never infer the molecular structure of an odoriferous compound solely by smelling it. Our sensory receptors and the neurons that integrate them *associate* molecular interactions with memories and prior experiences. The sensory

---

[10] Parr et al., "Perceived Minerality"; Parr et al., "Minerality in Wine."

experience of striking a flint becomes associated with an analogous sensory experience of drinking a nice Chablis, someone calls the Chablis flinty, and others then learn to call the same characteristic flinty even when they've never personally struck a flint. Language and flavor are both social experiences. We learn how to name flavors by listening to what others say.

I like minerality. I like the word, and I like wines with lots of it. If I say that the grenache blanc I'm drinking has great minerality, I hope that you'll hear something about the experience of drinking it. Saying that it tastes of Meyer lemon zest, greengages, and kelp-infused ocean air might raise more confusion—or, worse, a silly search for specific flavors and a sense of disappointment if you don't find them. I'd rather gesture to a mood than pretend that experiencing wine involves the kind of reductionist granularity that listing flavors is inclined to suggest.

Every day, we all make sense of new experiences by making quiet little arguments about their resemblance to prior ones. When I walk into a room I've never seen before, I probably won't struggle to find names for the furniture, even if the chairs aren't exactly the same as any chairs I've seen before.[11] I'll see that some objects are chair-ish— having something in common with objects I've previously learned to call chairs—and if I need to sit, I'll sit on those chair-ish things instead of on the things that look table-ish or credenza-ish. Such analogy-based thought tools make navigating the world possible, especially when everyone in the room uses "chair" to describe the same chair-ish things. We learn what words to attach to observations and experiences from the people around us, and in so doing, we learn which ones to think of as discrete and worth naming, in ways that curve back on themselves to shape how we experience things in the first place.

When you first began enjoying wine, you probably didn't taste minerality, or orange blossoms, or asparagus. By learning shared,

---

[11] This is Frederich Nietzsche's example, in an essay fantastically titled "Truth and Falsity in an Extra-Moral Sense."

socially sensible descriptors, the initially incomprehensible mishmash that follows the act of inserting wine in your mouth resolves into understanding. You experience minerality—really, actually—because you've learned to associate a particular sensation with a meaningful word, no molecular explanation required. Do minerally wines *actually* taste minerally? That's the wrong question.

5
Climate ✎

Cool-climate wines are more refined than warm-climate one. Their flavors are less accessible and require more finely honed sensibilities to enjoy. They're also more challenging to make, which obviously makes them better. These are absurd and patently unreasonable generalizations, but they don't come from nowhere. In a Portuguese study, casual drinkers systematically preferred riper wines, choosing a tinto from Lisboa over a pinot noir from Burgundy, for example, and tending to describe cool-climate wines as "unfamiliar." Wine students, in contrast, preferred the "difficult" cool-climate wines over the "easy" options favored by the hoi polloi.[1]

This study interrogated drinkers from sunny Portugal, so we should ask whether the same protocol repeated in chilly Alsace would find locals balking at Portuguese wines. Yet a French study comparing gamay wines from the coolest vineyards in the Côtes d'Auvergne with counterparts from warmer parts of Beaujolais found that French wine professionals preferred the peppery cool-climate wines; in contrast, untrained French consumers preferred the fruitier, less astringent, warmer-climate options.[2] And a study of Australian and New Zealand bottlings found that cool-climate wines steadily garnered higher ratings and higher prices.[3]

We could talk about rarity and popularity, whether having taste means liking things that most people don't, or what we might call the Oscar Wilde proposition, after his dictum that "everything popular is wrong." Any of those conversations, however, would elide a more fundamental question: how do you define a "cool-climate" wine region?

That question is stickier than it might appear because no one benchmark determines whose climate is in and whose is out. Here are nine

[1] Antunes, "Consumer Preference."
[2] Geffroy et al., "Impact of Winemaking Techniques."
[3] Schamel and Anderson, "Wine Quality."

possible ways to draw the line—though, in the end, desire and community may have more to do with the question than any line does.

Growing degree days. Many fruiting plants only grow when the ambient temperature rises above some baseline. For grapevines, that baseline is about 50°F. Growing degree days measure how much growing time a location affords. For each day of the year, add the daily high and daily low temperatures. Divide the result by 2 and subtract 50. Of those numbers, sum up the *positive* ones. (Discard the negative ones; cold days don't count in this system.) The result is the annual growing degree days for your location—which, confusingly, is a number of hours and not a number of days at all.

The Californian viticulturist A. J. Winkler, who popularized this system in the 1930s, judged that cool-climate regions have relatively few growing degree days.[4] That definition has the advantage of being both simple and quantitative. The numbers say you're cool, or they don't. The disadvantage is that those numbers aren't a great proxy for the things wine people care about: what growers need to do in the vineyard, and how the resulting wine tastes. Winkler's successors found in the 1980s that Carignan might take 935 degree days to ripen in one location and 1,582 degree days to reach exactly the same sugar concentration in another.[5]

Other ways to add up temperature. Which is why alternatives have cropped up to replace the growing degree day. The growing degree day's simplicity is also its weakness: it works from average daily temperatures and assumes that all heat, at any time, has an equal effect. Common sense says that's not right. A steady 65°F all day is nothing like a high of 95°F in the shade at lunchtime that becomes 35°F when night falls, yet in growing degree day terms, they're equivalent. Even more problematic at present is that this system doesn't account for days when ripening slows or stops because vines are too hot. Grapevines don't photosynthesize well above about 90–95°F. No photosynthesis, no sugar coming in, no ripening. That's one reason why ripening seems to require more heat in hot locations than in cooler ones according to the growing degree day system; the hottest portion of that heat counts in the model but doesn't count for the plant.

---

[4] Amerine and Winkler, "Composition and Quality."
[5] Mc Intyre, Kliewer, and Lider, "Some Limitations."

Numerous alternatives have been devised to more precisely get at a plant's lived experience. Some require measuring ambient air temperature throughout the day and adding up degree hours through the growing season—trivial with today's real-time digital weather sensors, though a good deal more resource-intensive when proposed in 1987.[6] Some account for the photosynthesis-dampening effect of chill and extreme heat. Some incorporate estimated day length by way of latitude.[7] All try to capture variability that matters to vine performance and wine quality, while avoiding extraneous detail that could cause data overload and make little difference in the end anyway.

Latitude. Latitude is a friend to New World wine marketers who can compare their vineyards' latitude with Fancy French Wine Region and say: "Hey, we're the same distance from the equator as folks who sell their wine for hundreds of dollars, so our bottles are a bargain for $34.99."

The problem, once again, is that latitude is too simple; this one number doesn't capture enough of what makes a place distinct. The 45th parallel north—halfway between the equator and the North Pole—transects Bordeaux, Piemonte, Oregon's Willamette Valley, Michigan's Leelenau Peninsula, the wine-growing south edge of Canada's Ontario, Croatia, and the southern part of Kazakhstan. None of these areas is hot, but they're all different. Brave, entrepreneurial souls in the Leelenau Peninsula are—slowly, and with increasing but variable success—working out how to vinify something other than cherries and sturdy European-American hybrid varieties. 🍇 Piemontese growers have to be careful about where they plant Nebbiolo—on warm hillsides—but they still succeed with this early-flowering, late-ripening grape responsible for intense reds from Barolo and Barbaresco. Meanwhile, Walla Walla, Washington is further north, at the 46th parallel, but inland, and known for hot summers that lead to full-bodied, often highly alcoholic reds. Oceans and lakes, altitude, and the finer aspects of regional topography have too great an effect on climate for sheer angle of the sun and day length alone to be singularly useful indicators.

---

[6] Mc Intyre, Kliewer, and Lider, "Some Limitations."
[7] Hall and Jones, "Spatial Analysis of Climate."

Viticultural concerns. The underlying problem with the options we've considered so far is that they aim for universality. In light of the substantial impact of local variables, universal rules fall short. An alternative is to stop asking quantitative questions of meteorological records and start asking qualitative questions of vineyard managers. Do you worry about your vines freezing in winter? Do you obsess over weather reports on spring evenings, hoping that a late frost won't interrupt bud break? Do you keep up with research from Cornell in upstate New York on cold-hardy rootstocks? If so, then you must be working in a cool climate.

Wine style. But maybe "cool climate" isn't about viticulture at all. As we've already seen, it's also shorthand for a set of wine styles. If you prefer transparent, fine-boned pinot noirs over full-bodied, brickish ones, you should be directed toward "cool-climate" wines. At that point, the designation has less to do with frosts or growing degree days and more with what a winemaker is trying to accomplish. Flavor variables may be constrained by climate, but they're also substantially shaped by human choices. For example, even though red grapes tend to accumulate less rotundone in warmer climes, this black-pepper-y compound can also be dampened by letting a lush leaf canopy shade ripening clusters, or by removing some skins before fermentation.

Fashion. Everyone with good taste prefers cool-climate wines, just as we all prefer dry wines and are only buying jammy red blends because our gauche friends want to drink them. That's obviously not true, but we can be more specific thanks in part to Greg Jones, who was once in the unusual position of being both an atmospheric scientist and the director of an undergraduate wine program in Oregon. Jones created a massive master list of wine scores by aggregating ratings from international wine critics, converting their various scales—100 points, 20 points, five stars, and so on—into one global rank order for each wine.[8] Then he looked for correlations between vintage and rank. Wines from warmer, dryer years consistently outranked wines from cooler, wetter ones. Vintages at the bottom of the list tended to come from short, cool seasons in which rain interrupted ripening.

[8] Jones, "Vintage Ratings."

Some especially hot years still brought forth average wines. But that's not the catch. Indeed, there are two catches. The first: Jones only examined wines from Burgundy, Bordeaux, Tuscany, and Portugal—all European, "Old World" regions. Do the same patterns hold for "New World" regions such as his own Willamette Valley, Australia's Barossa, or Chile's Valle de Maipo? That's a separate question because their climates, styles, and traditions differ. For example, Jones's research in Oregon demonstrates that daytime temperatures tend to be higher there than in comparable European regions, but nighttime temperatures tend to be colder, so while vines may flower around the same date, Oregonians harvest their grapes two weeks later.[9] Moreover, European grapes that aspire to becoming place-designated wines may be legally obliged, by regional regulation, to submit to harvest on a particular date or at a particular degree of ripeness. Grapes outside of Europe are rarely so constrained.

The second catch: wine ratings tend to be systematically biased toward "big" wines, whereas lither wines can be more drinkable over a whole glass or two. ⸱⸱⸱ Tasting is a completely different use-case from drinking, and professional tasting differs from casual tasting. Big (alcoholic, phenolic, super-ripe) wines make bigger impressions when tasters are faced with whole panels of similar wines at a time; they give tired senses something bold to sense. They're also inclined to wipe out the competition. A clever study from California found that when a trained tasting panel was given eight modest cabernets, each clocking in under 14 percent ABV, they described each as distinctly different from the higher-alcohol, 14–16 percent ABV cabernets they were asked to taste subsequently.[10] But when the same panel was given the higher-alcohol wines *before* the lower-alcohol group, they rated all of the wines across both groups as being very similar—even with mandatory pauses and bites of cracker between wines. The alcoholic wines overwhelmed tasters' ability to be discriminating about their lighter neighbors. Jones's ranking study only reports on tasting, not drinking—the use-case that interests most consumers at the end of

---

[9] Jones, "Spatial Variability."
[10] King, Dunn, and Heymann, "The Influence of Alcohol."

the day, unless your enjoyment of a wine derives principally from the number of points attached to the outside of the bottle.

Attitude toward global heating. Remember those brave winemakers on Michigan's Upper Peninsula? They've been replacing American native *Vitis labrusca* vines and European-American hybrids with more lucrative but less cold-hardy European *Vitis vinifera* varieties, to the tune of a 300 percent increase in just the first decade of the twenty-first century.[11] In the 1960s, Michigan's vineyards saw about 1,400 growing degree days per year (an imperfect measure, but it's something); between 1980 and 2011, the average increased to 1,628. That number has only continued to rise, so much so that geographers at Michigan State University predict that the state won't be "cool-climate" for much longer. Cotnari, an area in northern Romania long known for white wines, has begun producing reds in the past few decades. Climate modelers in Edinburgh and Copenhagen have suggested that Scotland may become a viable viticultural prospect in the foreseeable future.[12] Nova Scotia already is.

If you're a cool-climate vintner, global heating may let you get more sleep; if you're in a warm climate, it may put you out of a job. We can all agree that the climate crisis is a crisis, but the situation presents a more positive business outlook for some than others.

Tradition. Is cool climate about temperature, or about style? About style, or about image? Wine styles aren't just about the raw materials you have, but about who you want to be, and who you might want to be is tangled up in history and tradition. Germany and Austria (or, rather, some parts of what are now Germany and Austria) have long been lands of cool-climate whites and reds. Wine scientists at Germany's Geisenheim University say that where twenty years ago they needed to think about frost prevention, they're now concerned about overripeness and sunburn. Maintaining their classic wine styles seems more important—not just to consumers, but to the many other people invested in the identity of the region—than whether the university has needed to change its research program to maintain those styles.

Membership. When I switched fields from microbiology to rhetoric of science, near as I could ascertain, the only thing that students and

[11] Schultze, Sabbatini, and Luo, "Effects of a Warming Trend."
[12] Dunn et al., "The Future Potential."

faculty in my new English department had in common was that everyone did something with texts. As "texts" might mean anything from fifteenth-century morality plays to maps of the Oregon Trail, contemporary Iranian poetry, transcripts of teens telling jokes, or politically polarized tweets, that was a mighty broad way to define a group.

I didn't have a better answer until years later, when I became friends with a geographer. Their answer? What makes someone a geographer is whether they think of themselves as a geographer. Disciplinary identity, they suggested, is more about your orientation and desire for group membership than about your favorite research questions or the names of your degrees.

That way of thinking has helped me make sense of academic disciplines and cool-climate wines alike. The latter are made by people who want to talk and think about cool-climate wines, not by people who add up their growing degree days and measure them against some universal standard. It's a club. That orientation, that sense of common identity, will manifest in the way club members think about viticultural problems, wine styles, and marketing images, and in the kinds of wines they produce.

None of that is to say that physical constraints aren't important or can be wished away. Someone who studies the astrophysics of black holes will struggle to join the geography club, and Sicilian winemakers off the south tip of Italy will struggle to join the cool-climate one. But the astrophysicist might still attend a geography conference to deliver a talk about the geography of outer space, if she wanted to hang out with geographers. And some Sicilian winemakers are choosing to join the cool-climateers with vineyards planted improbably high up the sides of Mt. Etna. Maybe the best way to answer the question "Are you a cool-climate wine region?" is by asking in turn, "Do you want to be?"

6

Weather 🌰

Climate happens over years and decades. Weather happens over hours, days, and sometimes mere minutes. Climate is part of *terroir*, of the defining characteristics of a wine-growing place. Is weather? If it is (and maybe even if it's not), what relationship does intervening in it have to the wine produced as a result?

Let's take hail as a case study. Contemporary science has an oddly long history of trying to stop it, and it's an ever-present threat to numerous winemaking regions. An ill-timed storm can rip this year's fruit off the vine, destroy the leaf canopy needed to ripen it, and damage shoots and trunks to affect next year's crop, too. Fortunately, in most places, vintage-wrecking storms are an occasional tragedy rather than a constant fear. Unfortunately, they seem to be happening more often. Doing something about that—about the hail, that is; doing something about the climate crisis is a different matter entirely—*might* be possible, though what has been tried is totally out of proportion to what we know about whether it works.

How Hail Happens

Hail happens when a thunderstorm with strong updrafts meets a supercooled cloud, in which liquid water droplets are surrounded by colder-than-freezing air. Supercooling happens when water that would otherwise freeze has nothing to start freezing around, such as a bit of dust that can serve as the nucleus for a new ice particle. When such a nucleation site becomes available, ice forms fast. Initial ice particles function as nucleation sites for even more supercooled water. Updrafts bounce those particles up and down inside the cloud, giving them opportunities to collect more ice. Eventually, the resulting ice pellets become large enough that the force of gravity pulling them down outweighs the force of the cloud's rising air current pulling them up. Ice falls, and some unlucky bit of terrain has hail. Meteorologists know all of that but still can't predict exactly where hail will fall, thanks to rapidly changing conditions inside storm clouds and limited tools for tracking intra-cloud environments.[1]

The part of France barely south of Bordeaux sees the worst hail in that country, according to the Association Nationale d'Etude et de Lutte contre les Fléaux Atmosphériques (ANELFA).[2] ANELFA also reports that hail intensity increased by 70 percent between 2009 and 1989, when consistent record-keeping began. But ANELFA isn't chiefly a monitoring operation. Its primary raison d'être is to develop "artificial weather modification aimed at suppressing hail." This might seem fantastical, and yet the agency is merely one participant in a long tradition of scientific efforts to control precipitation. Nearly fifty countries currently mobilize anti-hail technologies, including France, Argentina, Australia, the United States, and China.[3] The strangest part of this story, however, is that no one seems able to confirm whether those technologies work.

In 1901, the journal *Science* carried a short report about "a special form of cannon which throws a large vortex ring at high velocity into the upper atmosphere," purporting to head off severe storms by prompting premature "collapse of an unstable atmosphere" (see the VORTEX RING box). The report's author, William S. Franklin of MIT, found this idea "absurd," but remarkably effective.[4] In a follow-up in 1924, Franklin softened his support, suggesting that "we will find that the word control is rather a strong term to use for what might be

---

[1] According to the US National Severe Storms Laboratory, research still hasn't explained exactly why some places are more hail-prone than others. Esteemed wine regions sometimes seem the hardest hit, but they're not. Russia, China, Wyoming, and northern Italy see some of the worst hail globally. Hail that doesn't hit anything of human value just doesn't make the news.

[2] The association's full name, fantastically, translates to something like the "National Association of Study and Struggle Against Atmospheric Scourges."

[3] Davitashvili et al., "Using Modern Technologies."

[4] Franklin, "Weather Control," 497.

VORTEX RING

You may well be asking yourself: what the heck is a vortex ring? First, what we're talking about is properly known as a toroidal vortex. Toroids are surfaces described by rotating a shape around an axis that the shape doesn't touch, which is a long-winded way of talking about a three-dimensional ring, like a donut. A vortex is a fluid swirling around a central point. Gases count as fluids for this purpose because gases can flow in the same way liquids do, so a tornado is a vortex, as is the contents of your wineglass when you give it a good swirl. Toroidal vortexes, it follows, are rings made of fluids that continuously swirl around the ring. You can make one by forcing gas out of a smallish opening in a biggish container, so that the gas at the edges of the opening curls back on itself. Smoke rings of the old-fashioned wizard-with-a-pipe kind are toroidal vortexes. Deployed at speed, they can be forceful enough to knock down light objects—or, as the occasion may require, disrupt "unstable atmospheres."

conceivably accomplished." Nevertheless, he remained keen on US experimentation with firing cannons into clouds.

Franklin wasn't a random eccentric whom *Nature* was indulging, but one voice among a hopeful chorus calling for practical experimentation. Though vortex cannons were indeed installed in some wine regions, they remained pretty clearly at the hoping stage until 1946, when a research program at the esteemed General Electric Company reported that dry ice and silver iodide could serve as artificial nucleation sites in supercooled clouds.[5] That report suggested that dropping bits of those materials into storm clouds could target rainfall to areas that needed it, and could prevent hail of significant size from forming at all—the first Rigorous, experimentally supported suggestion about how weather control should work.

[5] Ball, "Shaping the Law"; Harper, "Climate Control."

Offensive Defense Technologies

The greatest optimism about this new possibility probably occurred within the US military, leaders of which feared that the Soviet military might be even more enthusiastic than they were. Farmers in arid parts of the western United States also wanted to get in on the action, with some pooling resources to invest in the earliest commercial silver iodide cannons in hopes of directing rain to their fields while keeping hail away. The results were just as scattered as precipitation forecasts tend to be. Cloud seeding *might* have reduced the strength and severity of rain storms over crops, though not hail. However, without a reliable picture of when and where precipitation would have happened in the absence of intervention, the success (or failure) of intervening couldn't be established.[6]

Inconclusive results didn't stop the US Department of Defense from developing and eventually deploying rain-seeding technology in an effort to disrupt enemy movements during the Vietnam War. Under the code name Operation Popeye, the Thailand-based 54th Weather Reconnaissance Squadron—ostensibly *monitoring* the weather—was veritably out to *modify* it, targeting rain to wash out roads and intercept enemy troops with persistent mud.[7] When that project was brought to public light by the *New York Times* in 1972, the application was challenged in Congress, and the resulting furor led the United States and the Soviet Union to mutually agree not to weaponize the weather in 1975. Knowing that Popeye hadn't unambiguously returned on its multimillion-dollar investment no doubt made it easier to acquiesce.

Predictably, like Teflon and the internet, what was originally a defense technology is now used to turn profits in private industry. Artificial snow machines that pad ski slopes run on essentially the same technology that the military used. Airports routinely employ a similar strategy to keep freezing fog off planes. (In these settings, highly concentrated seeding treatments result in predictable, easily seen outcomes—very unlike diffuse treatments of massive, unpredictable

---

[6] Dessens et al., "Hail Prevention."
[7] Darack, "Weaponizing Weather."

storm clouds.) And, in the converse effort to try to make the sky *stop*
falling, wine industries remain major players.

Hopes for Controlling Hail over Wine Regions

In 2015, the nation of Georgia built an extensive anti-hail system
consisting of a radar detection system and a "missile strike" defense
array, comprising 85 silver iodide "rocket launchers" positioned every
10 kilometers across 650 acres of the vine-covered region of Kakheti.[8]
The rationale for seeding with silver iodide specifically is that this compound, also essential in some forms of old-fashioned photography,
resembles the structure of ice and so encourages water to make ice-like
patterns. It's also easy for landowners to generate by burning an ace-
tone silver iodide solution in ground-based generators. Still, we're not
exactly describing a minor intervention. In Kakheti, 60 mm rockets
carry silver iodide 2.5–4.5 kilometers up, releasing their loads over
about 30 seconds. Delta, the Georgian company responsible for the
technology, estimates that protecting Kakheti requires firing 5,000 to
6,000 such rockets each year. On a human scale, that can feel big; lined
up against the scale of a full-blown storm, it's not.

Kakehti's 2015 anti-hail network was merely an update; the original,
Soviet-built system was installed in the 1960s but not maintained after
1989. And Kakheti's story isn't an outlier. ANELFA manages something on the order of 650 silver iodide generators across France, including Bordeaux and parts of Provence. In Mendoza, the most prolific
wine region in Argentina and among the most hail-prone in South
America, the first cloud-seeding operation was up and running in
1959. Its results were inconclusive, and the same must be said of every
one of the six or seven technical updates Mendoza grape-growers have
seen in the intervening years. All the same, in 2021, winegrowers in the
Saint-Émilion region of Bordeaux cooperated to build a brand-new
cloud-seeding network relying on balloons to release seeding material. In this last case, contemporary innovation has been about better

---

[8] Davitashvili et al., "Using Modern Technologies."

predictions so that balloons are released at the optimal time. Cloud seeding doesn't have a chance of being protective if you don't seed the right clouds.

Generating robust scientific evidence about the efficacy of these expensive, extensive efforts is tricky because—who would have guessed—weather is unpredictable. In one of the earliest full-scale tests conducted by General Electric in 1947, seeding 180 pounds of dry ice into a hurricane "succeeded" at diverting its course, but the storm might just as well have changed course anyway.[9] Weather events can't be set up in identical replicates with an unmanipulated control. Hail is idiosyncratic, so conducting a standard sort of study to compare treated and untreated regions isn't an option; no one can outline two bits of land that they can guarantee will see equal amounts of hail before it happens, so establishing how much hail *would* have fallen in the absence of an intervention is impossible. Further complicating the pursuit of scientific certainty, commercial operations are understandably disinclined to conduct randomized trials if clients expect them to take action against every potentially adverse event, and we can only assume that research conducted in the name of national defense isn't been freely shared.

An expert group convened by the US National Academy of Sciences reported that the available experimental evidence arrives at a single inconclusive conclusion:

> Weather modification research has been in a state of decline in the United States for more than two decades. The reasons are many and include the lack of scientifically demonstrable success in modification experiments, extravagant claims, attendant unrealistic expectations (e.g., pressure from agencies to meet short-term operational needs rather than to achieve long-term scientific understanding), growing environmental concerns, and economic and legal factors. Within this context it became difficult to distinguish legitimate and important research from some cloud-seeding programs claiming success with little or no substantiation....

---

[9] Darack, "Weaponizing Weather."

We know that human activities can affect the weather, and we know that seeding will cause some changes to a cloud. However, we still are unable to translate these induced changes into verifiable changes in rainfall, hail fall, and snowfall on the ground, or to employ methods that produce credible, repeatable changes in precipitation.[10]

When the National Academy of Sciences convenes a working group, that group has strict instructions about its remit. Remits are defined not by scientific experts in the group, but by their funding. If a working group is charged with writing a report on weather control, that group will draw conclusions about the effects, risks, and benefits of weather control; it won't report on other responses to extreme weather. That requirement ensures that a group's problem is well defined and heads off arguments about misdirecting public funds. It also institutionalizes a major pitfall in talking about a technology: describing whatever technology you're investigating as the best, only, or most obvious solution, even when a better solution may involve doing something else entirely.

Winemakers have a different remit: they care about results, not report-writing, and about deliciousness, not national defense. By those measures, a protective, reactive response that doesn't involve advancing a technology development agenda may be better than a violent, preemptive one. On-the-ground protection against hail, such as netting, is a far more predictable response to that unpredictable threat than in-the-air prevention. Hail nets are expensive. They can reduce the sunlight that reaches vines—not negative everywhere, but an adverse effect in places such as Burgundy that tend toward overcast. Hail nets are also demonstrably effective.

Nets raise their own questions. Do you leave them up all the time, or rush to raise them when storms threaten? How valuable do grapes need to be and how often does hail have to fall for the expense to pay off? Robotic arms are making it possible to extend nets over rows when

---

[10] National Research Council, "Critical Issues," 68.

and only when weather is imminent, though that hardly helps with cost-benefit ratios.

Cost aside, it's difficult to imagine anyone objecting to hail nets on the grounds of their being an unreasonable intervention or a circumvention of *terroir*. Aggressively sending up rockets seems so much more intervention-y than defensively rolling out plastic mesh. But hail, and weather mitigation generally, is an excellent reminder that the line between unwarranted intervention and processes that are necessary to winemaking is fuzzy and subject to change across contexts. Hail may be a natural part of a landscape, but may also need to be excluded, rewritten, or canceled out to ensure that a vintage ends up in bottles instead of smashed in the vineyard. Erecting nets and seeding clouds would seem to be *structurally* no different from adding water to juice or removing sensory defects from finished wine when the choice is intervening and having drinkable wine versus not intervening and having nothing drinkable at all.

I don't want to end this chapter with the impression that non-violent responses to adverse weather events are somehow superior to aggressive ones. So before we leave weather behind, let's take a quick tour through a weather-related event that unquestionably calls for aggression: wildfires.

Smoke Taint

Firefighting is becoming a winemaking strategy—sometimes directly, as when California's 2020 Glass Incident Fire ate up significant bits of the Napa Valley. (Poor hail-ridden Kakheti was threatened by wildfires in 2021.) Vineyards themselves are remarkably fire-resistant; winery buildings, not so much.[11] More often, fire is an indirect problem by way of smoke and the smoke-tainted grapes that smoke leaves behind.

---

[11] Because vines are planted relatively far apart and vineyards don't otherwise host much brushy biomass, they burn so poorly that they sometimes even function as firebreaks.

Smoke taint is what it sounds like: an undesirable bonfire quality that manifests as a mouthful of ashes at its worst and dampens other flavors in minor cases. Unlike inhaling smoke, imbibing smoke residue isn't hazardous in the quantities we're discussing, so the worst thing about smoke taint isn't its (non-existent) health effects; it's that it plays hide-and-seek. The phenolics responsible for smoky aromas tend to tether themselves to grape sugars. Those sugars literally weigh them down, anchoring might-otherwise-be-aroma compounds in liquid solution so that they can't evaporate up into your nose. During fermentation—indeed, even during the very act of tasting, thanks to digestive enzymes in saliva—those tethers break, revealing smokiness where none may have been detectable before. ♪

Smoke taint's molecular offenders overlap with sensory compounds that winemakers sometimes add intentionally by aging wine in heavily toasted barrels. 🛢 Shiraz can also feature phenolics that head in a smoky direction. Clearly, some drinkers enjoy those flavors in moderation, so a slightly smoky wine may find a market. But more than the merest whiff of bonfire—especially in a wine style that can't pretend that it was supposed to taste that way—requires either aggressively removing the taint or aggressively removing the wine from this year's sales list.[12]

Removing all smoky molecules and only smoky molecules is difficult, between being hard to sniff out and overlapping with desirable flavors that one wants to keep. The least intensive option is to minimize how many smoky phenolics make it into the wine in the first place by minimizing contact between grape insides and grape outsides. Smoke compounds concentrate in skins. Unfortunately, so do color compounds, making this option no option at all for reds that derive their redness from fermenting juice on skins. For whites, bunches must be carefully hand-picked to avoid any in-the-field squashing that

---

[12] Or, for wineries that purchase grapes, aggressively refusing to accept tainted grapes from their growers. Breaking contracts is bad for everyone, so a major thrust of current smoke-related research is predicting whether or not grapes have been affected *before* harvest. This is tricky because smoke in the vineyard doesn't necessarily mean smoke in the glass. The timing of exposure matters, as does the type of grape and even the composition of the smoke.

might otherwise be unproblematic; even then, this mitigation method is only ever partially effective. Some winemakers may turn plans for reds into rosé, or may dilute a little bit of tainted juice in a big vat of something unaffected. Treating juice with activated carbon—the same stuff you might take if you've consumed poison, to help soak it up before your body absorbs the full dose—can soak up smoky phenolics, but comes with the unfortunate side effect of soaking up other, delicious phenolics that make wine taste like wine.

If money and equipment aren't at issue, you might turn to flash détente (literally "flash relaxation"), a technique invented in France in the 1990s to bolster color and tannin extraction and used widely for that purpose, but recently deployed against smoke taint. Flash détente is high-speed pasteurization 🐟 by way of reverse pressure-cooking: grapes are heated, then moved into a vacuum chamber where they cool rapidly and more or less explode. You've witnessed a similar effect if you've rapidly released the pressure from your Instant Pot only to find that the beans you hurried along have turned inside-out.[13] In addition to reducing the green characteristics of underripe grapes—a reason why it's popular in France—flashing grapes seems to mitigate smokiness, too.

Reverse osmosis, or RO, is a less dramatic maneuver. Effectively a molecular-level filtration process, RO allows winemakers to remove classes of molecules on the basis of how large they are, or to "tune" them by first removing them in their entirety and then adding some fraction back. As fantastic as that sounds, RO comes with major drawbacks. The method is good for smallish molecules, such as ethanol. A filter with smallish holes will let ethanol pass through while retaining larger molecular contributors to flavor and aroma; the ethanol "permeate" can then be selectively added back into the retained wine to tune alcohol concentration. ⋆⋆ Infelicitously, smoke compounds, especially when bound to sugar molecules as they often are, are big. A filter with holes big enough to let them pass is porous enough to let

---

[13] For those whose frame of reference is either before or after the early twenty-first century, an Instant Pot is a wildly popular electric pressure cooker/multifunction cooking appliance.

through nearly everything that makes wine wine, which is no help at all.[14] An RO filter also can't distinguish smoky compounds from their similarly sized, desirable molecular cousins. Processes that separate molecules by size aren't smart enough to distinguish things you want from things you don't.

Is This an Intervention?

Those are physical challenges. RO also comes with a philosophical challenge, or maybe it's a moral one, about intervention. Professing a "minimal intervention" philosophy has become The Thing To Do in some markets, with the idea being that humans should do as little as possible to grapes in the interest of letting terroir shine through. Detractors point out that wine doesn't make itself, so intervention must always be defined relative to some baseline of absolutely indispensable practice. For example, in theory, crushing isn't essential because grapes will break down of their own accord with time, pressure, and being moved around. Still, most winemakers would characterize crushing as an ordinary part of winemaking, not an "intervention." Other strategies, such as RO, may be indispensable in making a particular wine drinkable, but may still garner consensus as a substantial intervention, whether or not one believes that it should therefore be avoided (see the ARGUMENTS box).

The rare minimal-interventionist has accepted smoky grapes, vinified them as usual, and sold them as an unfiltered, unmitigated reflection of the year's flaming *terroir*. As I'm writing this, late in September 2021, Burgundy is looking back at a year of record-setting hail and reporting what might be their worst vintage in recent memory, with yields across the region at something like 10 to 30 percent of what they should be. Perhaps they could follow these bold wildfire-affected winemakers' example by selling empty bottles.

---

[14] Trying to remove smaller molecules doesn't always work the way you might expect on the basis of their molecular size because smaller molecules often form larger complexes that are too big and stable to pass through a filter.

ARGUMENTS

I see two arguments against interventions in winemaking, generally. The first is about boredom. The second is about systems.

Interventions, however defined, are a matter of making wine less like what you have and more like what you might want it to be. I see no moral argument against that; we send produce through all manner of manipulations to make wholesome and delicious drinks and foods, and lines drawn between morally acceptable and unacceptable manipulations can't help but be on shaky logical ground. The trouble is that manipulations of any kind take wine further away from produce and closer to following a recipe through which it becomes possible to standardize flavors. Standardization is welcome in many processed goods, including big-brand wines that sell on offering consistency from bottle to bottle. In contrast, working with what you have may sometimes mean suboptimal flavors—but what the heck is an optimal flavor? For my own part, I would rather have every wine I drink be different, even if not every bottle is earth-shatteringly wonderful, than have endless access to some version of "the best" defined as what's most likely to sell in the highest volumes. Then again, I drink wine; I don't sell it.

The difference between boredom and interest obviously isn't the difference between intervention and non-intervention. Empty bottles are boring, even when shipped from hail-shredded vineyards. RO and myriad other overtly manipulative winemaking tools can be applied in service of an idiosyncratic vision. So we can't just talk about interventionist versus minimal-interventionist philosophies. We have to draw distinctions on the basis of how interventions are used and to what ends.

The second rationale I see against intervention has to do with systems and what we don't understand about them. Vineyards are complex systems. So are weather systems, so much so that meteorologists still can't predict where hail will fall. And so is wine itself. Intervening in a complex system always carries the risk—more like the inevitability—of changing things that you didn't mean to change. Controlling the weather to avert hail from your vineyard or redirect rain away from your wedding day sounds great, save that even if we knew that weather modification worked, we still wouldn't know what we're doing. We have too many examples of unintended consequences following from technologies applied through ignorance, hubris, and haste for me to be enthusiastic about inviting more of the same.

7

Yeast ✺

Saccharomyces cerevisiae, our old friend brewer's, baker's, or winemaker's yeast, earns its keep in kitchens and wineries by being unusually proficient at converting sugar into ethanol and carbon dioxide. But yeast is more than an alcohol factory. It's a companion that humans have evolved alongside. We've become accustomed to caring for and working with each other. And among its other awesome powers, yeast is good at prompting humans to consider our assumptions about other-than-humans—even, maybe especially, about, the smallest ones—precisely because yeast can be a process, product, factory, and friend all at the same time.

What Is "Yeast"?

Before we continue, a note on terminology: "yeast," nineteen times out of twenty, means *Saccharomyces cerevisiae* (see Figure 7.1). The twentieth time, someone is talking about a yeast infection caused by *Candida albicans*, a species completely unrelated to *Saccharomyces*. The world of yeast and fungi is vast, its own kingdom, and we tend to use species names for its other members.

Biologically, yeast and fungi are the same kind of critter, but by convention, "yeast" is attached to those that live primarily as single cells, while "fungi" is reserved for primarily multicellular organisms. Some can switch between these lifestyles. That's why the genus of yeasts commonly known as *Brettanomyces* has a second name, *Dekkera*. ✺ The same microbe, genetically speaking, is *Brettanomyces* when living as single cells and *Dekkera* when hanging together as a group.

But what is *Saccharomyces cerevisiae*, exactly? This question is trickier than you might expect for a creature we know so well. The

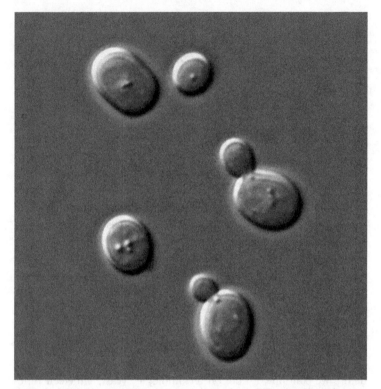

Figure 7.1 *Saccharomyces cerevisiae* cells under a differential interference contrast microscope, which amps up contrast in structures that are usually more or less transparent, as these cells are. Several of the cells are budding, splitting off a bit of cytoplasm to form a new cell. Image by Masur, in the public domain.

thing is that the idea of "species" as a biological category isn't apt for microorganisms in general. Whereas macrobes like us have one characteristic set of DNA, one genome that distinguishes us for life and that we pass to our offspring, many microbes pick up DNA from their environments, so one microbial lineage can't be counted on to retain one stable genetic identity over generations. *Saccharomyces* are especially genetically promiscuous. As a result, the taxonomic lines separating closely related species within the *Saccharomyces* genus— *S. cerevisiae, S. bayanus, S. pastorianus, S. carlsbergensis,* and so on—have been redrawn over and over as an expanding range of strains are found and examined with new tools.

Humans have working relationships with other *Saccharomyces*, especially in brewing beer, but our business in winemaking is with *S. cerevisiae*. It's easy to think of that relationship as one-sided and hierarchical, with humans domesticating yeast for human benefit. But if we take a more expansive view toward other creatures, it seems more appropriate to say that yeast and humans have co-domesticated each other.

Co-domestication

Domestication is often imagined as an event initiated by humans, but other creatures actively participate too. Evolutionary anthropologists suspect that gray wolves made the first move in the chain of events that led to domestic dogs. The hypothesis goes that some wild canines were initially attracted to food and protections associated with human habitations. Over time, those bold individuals and their offspring adapted to anthropogenic niches, growing culturally and genetically distant from their less sociable cousins. Because gray wolves ranged so widely across the Northern Hemisphere, identifying just one time and place where dogs first broke off from the pack hasn't yet been possible, but we know that dogs were firmly established in human households by the end of the Paleolithic period, about 11,700 years ago.[1]

Not all now-domesticated animals initiated their lifestyle changes, certainly. We know that humans made deliberate efforts with shier critters: rabbits, bees, and horses, for example. Still, even these relationships have been reciprocal: these creatures sculpt human habits to their needs just as humans sculpt their habits to ours. But in the earliest cases—first dogs, soon thereafter cats, chickens, ducks, and carp—that reciprocity is so strong that we should say that other-than-human and human creatures have domesticated each other in an ongoing process driven by mutual benefit, requiring mutual adaptation, and leading to lifestyle changes on all sides.

---

[1] Larson and Fuller, "Evolution of Animal Domestication."

Reciprocal domestication applies to *creatures*, not just animals. Corn is among the most successful species on the planet because it's succeeded so magnificently in seducing humans with sugar and starch.[2] Yeast, in its own way, has done the same with alcohol and carbon dioxide.

Yeast is more like dogs and chickens than rabbits or bees: they invited themselves into human domiciles rather than needing to be deliberately wooed. Though, as with dogs, paleogenetics has yet to pinpoint precisely how early wine yeasts and human fermentations came together, since their predecessors were so widespread. The ancestors of present-day wine yeasts genetically resemble strains that live on tree sap in Chinese forests. Yeast may have been first domesticated in China before migrating to Europe, or wild Chinese yeasts may have found their way into Europe and been domesticated there, concurrent with separate domestication events back in Asia.[3] What we do know is that yeast and humans have been friends for at least as long as dogs and humans and probably longer.

Until very recently, archaeologists have thought that Neolithic people—the folks socializing with not-quite-wolves—subsisted mostly on meat. Starchy plants didn't seem to be a major part of the diet, a story supported by piles of animal bones at places where Neolithic people partied, like Göbekli Tepe in present-day Turkey. That story has fueled the contemporary high-fat, high-protein, grain-free paleo diet craze. But it seems to be incomplete. Large stones at Göbekli Tepe, previously thought unimportant, show signs on further examination of having been used, extensively, for grinding and stewing enormous vats of porridge and beer.[4] The two were probably similar: both a kind of sludge, both spontaneously fermented, both made from grains just beginning to be domesticated at the time.

A salient factor in making present-day sense of Neolithic diets is that fossilized bones are far easier to find than fossilized fragments of bread. Finding fragments of ancient leafy foods or microbes is even more difficult. Yeast bodies don't fossilize well, so their presence must be inferred from telltale chemical residues. When the insides of ancient clay jars are scrutinized with sophisticated microscopy, archaeologists find conclusive evidence that alcoholic beverages are ancient. Chinese villagers were

---

[2] A case that Michael Pollan presents in *The Omnivore's Dilemma*.
[3] Steensels et al., "Domestication of Industrial Microbes."
[4] Curry, "How Ancient People Fell in Love."

enjoying rice-based alcoholic drinks by 7000 BC. Early versions of wine were well established around the Zagros Mountains by 5000–6000 BC.[5]

   We may not yet know the route yeast and wine took from one side of the Caucasus to the other, but it's easy to imagine how yeast would have enjoyed the hospitable environment provided by human-made stores of something sweet. Then, prehistoric people would have sustained relationships with particularly delicious yeasts by learning what kinds of environmental conditions they liked best, or by backslopping, using a bit of one favorable ferment to inoculate a new batch. Yeasts that thrived on the plentiful resources humans could provide would become those most adept at producing deliciousness, and would grow genetically distinct from those that weren't. Humans learned to maintain microbes that made food taste good and to rely on their produce, all millennia before Pasteur conclusively settled the debate about whether yeast was responsible for fermentation.

Adapting Together and Adapting Apart

Humans have metabolically adapted to digesting alcohol by way of two specialized enzymes: first, alcohol dehydrogenase degrades ethanol into acetaldehyde; then, aldehyde dehydrogenase degrades acetaldehyde into acetic acid (see Figure 7.2). Acetic acid is a common metabolic intermediate, easily digested into carbon dioxide and water. However, both alcohol and acetaldehyde are toxic if left floating about in one's cells or bloodstream. The degree to which that happens varies

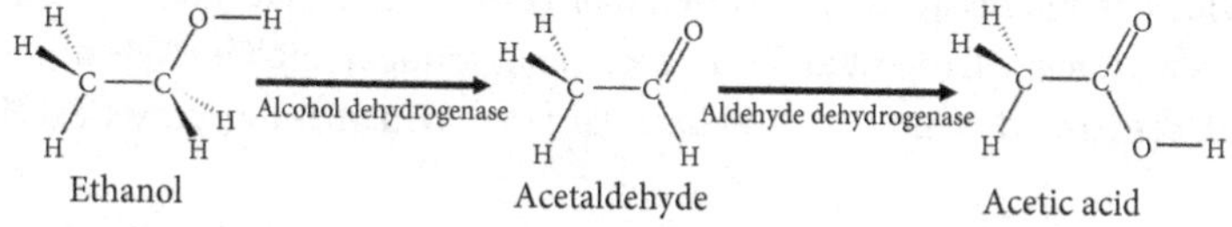

Figure 7.2 The two-enzyme pathway primarily responsible for ethanol degradation in humans (*Saccharomyces cerevisiae* can use an effectively identical pathway).

---

[5] McGovern et al., "Early Neolithic Wine." Patrick McGovern has worked extensively on ancient ferments as scientific director of the Biomolecular Archaeology Laboratory for Cuisine, Fermented Beverages, and Health at the University of Pennsylvania Museum.

from person to person, even when they've consumed equivalent quantities of alcohol, because individuals produce different quantities of these two enzymes. Members of some Chinese, Japanese, and Korean lineages make an extremely efficient version of the first enzyme that produces acetaldehyde so quickly that their comparatively less efficient version of the second enzyme can't keep up. The resulting acetaldehyde pileup is responsible for the flushed face, headache, and other discomforts—collectively termed acute alcohol sensitivity—that often lead people with this metabolic configuration to avoid drinking.

After innumerable generations of adapting to particular human-driven fermentation jobs, domestic yeast populations—behind beer, wine, bread, sake, coffee, chocolate, and so on—have become genetically distinct from each other as well as from their forest-dwelling counterparts.[6] Genetic resemblances make it fairly easy to trace family trees. Genome sequencing shows, for instance, that the US brewing industry has grown up with yeasts resembling those used in the United Kingdom, not yeasts found in North American forests, making it quite clear that beer-loving colonizers brought their microbial friends with them.[7] Yeasts that process coffee and cacao seem to have developed from genetic exchanges among European wine yeast and North American and Asian forest yeasts, with patterns that map remarkably well to early human movements among those places.[8]

Domesticated animals and plants have signature characteristics that distinguish them from comparable non-domesticated creatures. Humans show this "domesticated phenotype" too, with physical and behavioral changes akin to those seen in other species. So do domesticated microbes, even if their differentiating markers aren't so obvious as smaller teeth and varied hair color. S. cerevisiae employed in brewing ale will digest maltotriose, a sugar found in malt but few other places; tree-dwelling S. cerevisiae cannot.[9] Winemaking yeasts excel

---

[6] Both coffee and chocolate begin as seeds surrounded by fleshy fruits. In most production methods, S. cerevisiae ferments the fruit off of the coffee bean or the cacao pod before the beans and pods themselves are processed.

[7] Steensels et al., "Domestication of Industrial Microbes."

[8] Goddard, "Microbiology."

[9] Maltotriose is also found in your mouth, as the product of salivary enzymes acting on starches as you chew, but you would know if you had S. cerevisiae dwelling in your mouth and you wouldn't be happy about it.

at withstanding the antimicrobial effects of sulfur dioxide because they produce unusually large numbers of a membrane transporter that shunts sulfur out of the cell. 🆂🆄 Sake yeasts produce generous quantities of metabolic byproducts that smell like apple and banana (ethyl caproate and isoamyl acetate, respectively). Yeasts used in molecular biology research are nutritionally easygoing and often especially ready to take up environmental DNA. Each professionalized group has its own specialized adaptations.

Genome for genome, wine yeasts resemble their undomesticated cousins more closely than most other domesticated groups. Considering their respective lifestyles helps explain why. Brewing yeast strains often move straight from one batch of beer to the next, batch after batch. After many generations of this pampering, they've lost much of their genetic capacity to adapt to environmental variation. Since they needn't survive beyond the warm, comfortable confines of a brewery, when random genetic mutations damage their ability to be more flexible and resilient, nothing changes; their lifestyle doesn't demand flexibility.

Wine yeast strains, in contrast, are seasonal workers. They're exposed to far more stress and variable conditions than those employed in brewing or scientific laboratories. They labor through acidic, perilously sweet, 🍷 wildly alcoholic conditions that would wipe out most microbes. Few of us are asked to do our jobs while scuba diving, but that's effectively what's demanded of yeast fermenting at the bottom of the gargantuan tanks used in high-volume wineries, where they experience the pressure of several tons of grape juice weighing down on them. Successful wine yeasts are highly specialized experts, the pearl divers among *Saccharomyces*.

Some of the most robust of these mission specialists have gone corporate, specially selected and deliberately developed by companies looking to deliver strains that tolerate more alcohol, more thoroughly outcompete other microbes, or produce (or avoid producing, as the case may be) specific flavorful metabolic byproducts. Having had their capabilities experimentally verified, these yeasts are grown in uniform monocultures and shipped around, often as dry granules in foil packets like the ones you might buy for home baking.

Inoculated Versus Uninoculated?

The wine world tends to distinguish ferments that begin with commercial yeasts from ferments that begin through the initiative of whatever ambient yeasts happen to move in. The thing to remember is that ambient yeasts are domesticated too. If we embrace the premise that humans and (wine and other) yeasts have co-evolved, adapting to each other's company, then we have to rethink the common distinction between wines fermented "naturally," with "wild" yeasts, and wines inoculated with cultured strains.

Wild winemaking yeasts, in the sense of yeasts untouched by human influence, don't exist. They'd be unlikely to make decent wine if they did, since a random yeast strain plucked from an untouched forest won't be adapted to tolerate extreme ethanol, acids, sulfur dioxide, and contemporary winemaking's other threats. Yeast strains responsible for non-inoculated ferments may be tenacious long-term denizens of particular winemaking locations, having developed alongside their industries. Some may have crossbred with commercial strains. Some may even *be* commercial strains that have moved out of their foil packets to walls, barrels, or other winery-associated residences. Conversely, the yeasts responsible for *inoculated* ferments aren't necessarily the commercial strain that emerged from the foil packet alone, because the hospitality of a vat of juice may be accepted by a winery's other microbial residents too. In short, natural winemaking and (microbially) cultured winemaking look more like an "and" proposition than like mutually exclusive choices, let alone mutually exclusive choices made by humans.

Commercial yeast as we now know it became available around World War II, when the need to feed soldiers pushed researchers to develop more shelf-stable, easily transportable, rapid means of raising bread. For centuries prior, home bakers had been sending children to fetch a pint of yeast from the local brewhouse in the form of froth from the top of a fermenting vat.[10] Before 1676 (and for quite some

---

[10] Brewhouse yeast made "sweet" loaves possible, in contrast to home-cultivated sourdoughs. The popularity of bottom-fermenting lager yeasts over top-fermenting ale yeasts spurred on commercial baking yeast innovation, since yeasty froth can't be collected from the latter.

SLUDGE

If frothy sludge doesn't sound very appealing to you, you're not alone. *S. cerevisiae* has historically been the one microbe that everyone can agree to love, but even it has attracted occasional haters. At the height of the fad for purity and hygiene in Victorian Britain, the kind of fermentation responsible for both bread and beer was construed as a wholly undesirable kind of decomposition or "putrefaction," at least by some members of the ultra-hygienic classes. Fermentation was difficult to control, smelled, and made alcohol. Dough weighed less after fermentation than it did when first mixed, which was taken as scientific evidence that yeast wasted part of the valuable, nutritious flour that could and should be put to better use. Consequently, in the eyes of social improvers out to maximize efficiency and enforce cleanliness, yeast-risen bread was a social ill that should be replaced by efficient, clean, chemically and mechanically produced loaves.

We have this era to thank for the spongy, tasteless, additive-packed rectangles that continue to line grocery store shelves masquerading as the staff of life. When first introduced, heavily processed white bread that literally has air beaten into it so that it rises faster was considered (by some Victorians, at least) *more* healthful than slow-raised sourdoughs that now fetch far higher wholesomeness marks. Yeast—destructive, wasteful, dirty, and uncontrolled—even became a metaphor for working classes who didn't subscribe to higher society's hygienic ideals and who couldn't be satisfactorily controlled by those who did. The nineteenth century, as lived by a small group of privileged people clinging to a damp island on the edge of the Atlantic, was one of the few moments in human history wherein humankind's smallest best friend has instead been considered an enemy.

We've mostly recovered from that idea, but some of today's health fads have people thinking that yeast is bad for them too. Conflating the *Saccharomyces* species that ferment wine, beer, and bread with the *Candida albicans* responsible for yeast infections is a common mistake, even though they're completely separate species, and consuming one doesn't put you at risk of the other. One of the eminent

> founders of *Saccharomyces* genetics, Ira Herskowitz, was badgered by a man who thought that the professor was simply being cruel for not responding to the man's insistent requests to help with his wife's debilitating yeast infections. Sadly, the man was barking up entirely the wrong tree.

time thereafter), when Antonie van Leeuwenhoek first reported seeing "little animals" under his microscope, the word "yeast" just meant froth, from an ancient Sanskrit word for "boil"—a fair description of the appearance of yeast in action (see the SLUDGE box). Shakespeare is responsible for an early recorded use of the word in English, in *The Winter's Tale*, first published in 1623, wherein a ship caught in a storm is precipitously rising up and sinking low on high waves, alternately "boring the moon with her main mast, and anon swallowed with yeast and froth, as you'd thrust a cork into a hogshead."

In England, the particular froth you might fetch from the brewery for baking day was *barm*. You might do that, even though both bread and beer will ferment on their own, because inoculating comes with a measure of reliability and produces distinct flavors. In his 1879 *Studies on Fermentation; The Diseases of Beer, Their Causes, and the Means of Preventing Them*, Pasteur observed that winemakers never added yeast but brewers always did, because beer is far more susceptible to injury from unwanted microbial growth than wine, and beginning with a big dose of yeast offered some protection.[11]

However, barm must be used fresh, before *S. cerevisiae* is overwhelmed with unsavory bacteria or molds happy to inhabit a moist, nutrient-rich environment. In pursuit of a longer shelf life, a patent for dried yeast, probably the first, was submitted in 1796. Many more followed, but the process always killed so many yeast cells that no dry product could approach the efficacy of fresh for more than a

---

[11] The English translation of Pasteur's seminal work was published in London in 1879 with this title. The original was published in French in 1876 under the title *Etudes sur la bière: Ses maladies, causes qui les provoquent, procédé pour la rendre inaltérable, avec une théorie nouvelle de la fermentation.*

century. The dry yeast that the Red Star company developed around World War II therefore initiated a revolution.[12] It was highly active, lightweight, and stable for months if not longer. Being prepared from a pure stock culture maintained by company microbiologists, it could claim to be the same yeast every time, not a variable mix of whatever strains happened to dominate a particular batch. From being the *process* of how fermenting happened, yeast became an *ingredient*.

Yeast became not just an ingredient but a "pure" ingredient, and not just in the sense of being unadulterated. In 1889, while employed by the Carlsberg brewing company in Copenhagen, Emil Christian Hansen had isolated the first individual yeast cell by suspending a drop of liquid from the bottom surface of a glass slide in a sealed chamber. Letting one and only one cell divide and keep on dividing yielded a genetically identical culture, the famous "Carlsberg bottom-fermenting yeast number 1."[13] Thereafter, propagating single cells became a strategy to ensure that commercial yeasts performed consistently.

Commercially standardized yeasts were especially welcome in North America and Australia, where winemakers sometimes blamed differences between their modest efforts and the greatness of established European houses on the ignobility of their local yeasts. Where European wineries had cultivated well-heeled yeasts over generations, so the story went, New World upstarts were dealing with microbial savages. (This sounds racist, and it was.) By the early 1980s, most American and Australian winemakers relied on commercial yeasts to conduct their fermentations. Bryce Rankine, an oenologist who defined the Australian wine research scene from the 1970s through the end of the twentieth century, argued that this practice was responsible for the widespread high quality of Australia's bottlings. Rankine thought that spontaneous fermentations worked well in France because the strains on grape skins in established French vineyards had been refined over generations of cycling between winery and vineyard.

---

[12] Red Star still sells yeast, though it merged with Lesaffre, another yeast supplier with a long pedigree, in 2001.

[13] Strictly speaking, propagating one yeast cell doesn't guarantee a genetically uniform culture, since individual cells can still suffer random mutations. However, those occurrences are sufficiently rare that they're routinely ignored.

Australian winemakers working with new vineyards, lacking the same cultural history, needed to inoculate scientifically selected strains.

Grape skins were thought to be yeasts' primary vineyard home until the 2000s, when a flurry of microbial ecology studies gradually established that *S. cerevisiae* cells are few and far between on the skins of intact grapes. This makes sense. Licking an intact grape will readily demonstrate that its outsides aren't sweet. On the other hand, sugar-loving *S. cerevisiae* propagates rapidly on damaged grapes whose sugar-rich interiors are exposed. So how do the *S. cerevisiae* on those grapes get there? A popular hypothesis implicates insects as vectors of microbial movement from reservoirs of yeast life on vine trunks, topsoil, and broken berries, which are of course as attractive to insects as to fungi. Insects are probably also responsible for moving microbes around wineries and between wineries and vineyards, along with humans and their equipment.

So, how to know which yeasts are *really* doing the work? Inoculating is no guarantee that the inoculated strain will conduct the bulk of fermentation. Not inoculating is no guarantee that commercial strains resident around the winery won't move in. The only way to be sure is to check. "Checking" means drawing off a bit of liquid from the ferment in question, sequencing whatever *Saccharomyces* DNA can be found in it, and comparing the results to DNA sequences from known industrial strains. None but the very largest wine producers will have that capacity in-house, but industry-facing research institutions and fee-for-service labs can run those tests. So can researchers trying to draw general conclusions about yeast populations at work across the industry, though their results suggest that how these things play out may differ from place to place.

In one case, researchers at the University of British Columbia checked on yeast working in one specific place: uninoculated pinot noir ferments at Stoneboat Vineyards in the province's southern Okanagan Valley. They identified a mix of yeasts: some obviously commercial, some unrelated to commercial strains, and some that looked like crossbreeds.[14] In another case, a Bordeaux-based group scrutinized

---

[14] Martiniuk et al., "Impact of Commercial Strain."

yeasts dwelling in several wineries in Sauternes and found that a mere 7 percent could be mapped to commercial strains. At least some had been resident in their winery homes for several decades and probably longer.[15] A third, extensive study of spontaneously fermenting New Zealand sauvignon blanc, turned up an insignificant number of commercial strains among nearly 300 genetically unique, non-commercial isolates—a number that, extrapolated, suggests that New Zealand might harbor as many as 1,700 unique wine yeasts. 🖐 Precisely in how they don't all agree with each other, these kinds of investigations are demonstrating that regions and even individual vineyards bear distinct populations of microbes—including fermentation-ready *S. cerevisiae*—that probably contribute to wine style. Also, commercial yeasts haven't achieved world domination.

Naturecultural Yeast

As in many collegial relationships, winemaking yeasts and winemaking humans work together in a spectrum of ways. Some humans dump enough sulfur dioxide into their juice to eliminate most microbes before establishing their one carefully chosen commercial strain. 🆘 Some leave juice to ferment as it will, driven entirely by local microbial populations. Plenty follow some kind of middle way, including contemporary variations on the ancient backslopping theme.

Microbial genetics is expanding options for the middle way. DNA sequencing now makes it easier to identify the yeast one is dealing with. Consequently, potentially unique autochthonous strains—strains *of their place*—can be isolated from vineyards or wineries, interviewed for their qualifications as cellar workers, grown up in a monoculture, and brought onto the winemaking team much as a winemaker might select a standard commercial strain, but with the added benefit of a good story and maybe an amplification of terroir. 🖐

Good stories aren't the only reason to pursue this fresh approach to microbial material. Several surveys of commercial wine yeasts

---

[15] Börlin et al., "Cellar-Associated *Saccharomcyes*."

have found that they're disappointingly similar, with strains sold by competing companies under different names being effectively identical.[16] That's obviously not great news for the companies, but it also isn't great for anyone else, since winemakers who inoculate seem to be making limited use of yeast's wide-ranging genetic diversity. Prospecting for new strains that have adapted to winemaking but haven't been commercially developed could expand the pool.

The wine world has seen a fair bit of hubbub around what to call these varied permutations of yeast-working. "Inoculated ferment" is uncontroversial. But calling the contrast case "wild" ignores our long history of co-domestication. Calling them "spontaneous" distinguishes between winemakers who do nothing to start fermentation and those who use some form of homegrown starter culture, which might not be the distinction one wants to make.[17] Calling ferments "natural" might be the least satisfactory of all, with one set of critics eager to observe that no competent wine yeast is remotely natural because wine yeasts have always already been cultured, and a different set equally eager to observe that plenty of non-inoculated ferments are a far cry from what they would call "natural wine." 🍷

In a way, this terminological soup sums up the whole history of winemaking as a multispecies activity. The story of wine yeast is a tale of how the world can't simply be carved up into natural and artificial, or into wild nature and human culture. Yeast aren't either commodities or creatures, processing agents or participants; they're all of those things. The world is a naturecultural place.[18] Human actions send ripples out into environments that might seem at first glance to be separate from human habitations. The wider, wilder world continually intrudes into human-built ones. "Pure" nature and "pure" culture are impossibilities; humans shape the more-than-human world, and it shapes us back.

---

[16] Borneman et al., "Whole Genome Comparison."

[17] The spontaneous crowd also includes winemakers who rely on temperature cues to communicate with microbes, chilling juice in the first instance and then warming it up when they want fermentation to begin, which isn't "doing nothing," but which also doesn't involve directly changing the microbial dynamics of the mix.

[18] And in the fantasy world in which I get into the winemaking business, I'd make naturecultural wine, not natural wine, to acknowledge that wine is always a dialogue between human ideas and the much-more-than-human networks through which we play them out, in ways that are never simply about either exercising or giving up control.

After more than a century of studying individual microbes in isolation, contemporary microbiology is studying how microbes live in complex communities, with each other, and with us, painting a more complex picture of a more-than-human world . . . which brings us to the next chapter.

8

Microbiome ✿

Louis Pasteur was clear on the difference between good microbes and bad ones. Yeast were good. Bacteria were bad.[1] While Pasteur may be best remembered for contributions to human health these days, he initially earned accolades for studying "diseases" of beer and wine.[2] In his 1866 *Études sur le vin; ses maladies, causes qui les provoquent, procédés nouveaux pour le conserver et pour le vieillir,* he identified bacteria as the cause of "alien ferments" other than alcoholic fermentation—the only kind that had any business in wine—and was confident that he'd thereby solved the industry's problems.[3] Having found bacteria that caused spoilage, he didn't know to look or, frankly, have any interest in looking for bacteria that didn't. Pasteur's accomplishments are incontrovertible, but he was wrong on this point.

Eliminating Bacteria

Pasteur was an excellent *applied* scientist, fundamentally interested in practical application over explanation. He simply didn't care about

[1] The wonderful world of microorganisms also includes creatures other than yeast and bacteria: microscopic animals such as planaria, and archaea—a whole wonky kingdom of creatures that superficially resemble bacteria but are as evolutionarily distant from them as humans are. I'm ignoring archaea for the rest of this chapter because there's not much to say—not because archaea aren't both important and insanely cool, but because so little has been done to understand how archaea and wine might intersect.

[2] Bruno Latour, a founding figure in my own field of science and technology studies, wrote an entire book about the "great man" problem using Pasteur as his case study. Latour's point was that Pasteur, as he's been constructed as a French national hero, is an invention, not a man. Even though innumerable other characters—including microbes themselves—contributed to Pasteur's success, the icon of Pasteur as a singular genius has been fabricated to support a particular, politically motivated story of French triumph. Pasteur is an obvious example for Latour, him being French and all, but Latour's point is about the problem of iconic figures in history in general. See Latour, *The Pasteurization of France.*

[3] The title translates as "Studies on wine; their diseases, causes that provoke them, new procedures for preserving them and aging them."

classifying microbial life beyond distinguishing useful microbes from spoilers and establishing protocols to kill the latter. Controlled heating—what we now know as pasteurization—was a perfectly satisfactory solution to the killing part in Pasteur's book, so as far as he was concerned, his work here was done.

Pasteurization may now be synonymous with milk, but Pasteur initially proposed it as a cure-all for the "filaments," hazes, putrid odors, and other diseases to which nineteenth-century wines were regrettably prone. (It *is* still used for wine, largely when overt bacterial contamination threatens to make a batch unsellable, though more precise "flash" technologies are expanding its range of applications.) 🐌 Pasteurization was a big hit with the French navy and with wine merchants who were keen to preserve marginal stock.[4] Quality-oriented folk were more reticent, despite substantial public relations work by Pasteur, other scientists, and the French ministry of agriculture, all of whom were adamant that heating didn't damage quality. Even if you believed them, though, the trouble remained that heating was difficult

---

HEAT

Heat-treating itself wasn't new. Nicolas Appert introduced France to a rudimentary form of canning in the 1810s, and heat as a mode of preservation pops up in records dating back to Roman times. Everyone therefore knew that boiling wine left it unappetizingly cooked. Pasteur's method was innovative because his experiments indicated that mere minutes at a modest 130–140°F (55–60°C) would do the antibacterial job. However, rapidly heating and cooling large volumes of liquid was a technical challenge, and one in which Pasteur himself had no interest; that, in his eyes, was work for technologists, not scientists.[5] The other issue was that Pasteur and the wine microbiologists who followed him were more concerned with "soundness," or the absence of microbial faults, than with flavor. Understandably so; microbial faults could be dire. Thanks to

---

[4] Paul, *Science, Vine and Wine.*
[5] Despite technical improvements, until recently, pasteurization couldn't avoid hot spots that would taint a wine with cooked flavors—flavors you'll recognize if you're

their predecessors' work, contemporary scientists have the luxury of tackling other priorities. Unfortunately, wine microbiology's early years also established an expectation that bugs are bad until proved innocent, which has arguably been less helpful.

to control (see the HEAT box). And when applied shortly after alcoholic fermentation completed, per the recommendations of Pasteur and company, it disrupted what we now call malolactic fermentation.[6]

Controlling Bacteria

Malolactic fermentation, or MLF, is a fairly recent name for an age-old process by which bacteria metabolize the malic acid found in grapes into lactic acid (see Figure 8.1). Malic acid tastes sharp and pointy like a green apple; apples belong to the genus *Malus*, after which malic acid is named. Lactic acid contributes to the gentler flavor profile of fermented milk products such as yogurt, which result from lactic acid bacteria digesting milk sugars; *lact-* is the Latin root for "milk." Lactic acid bacteria is a

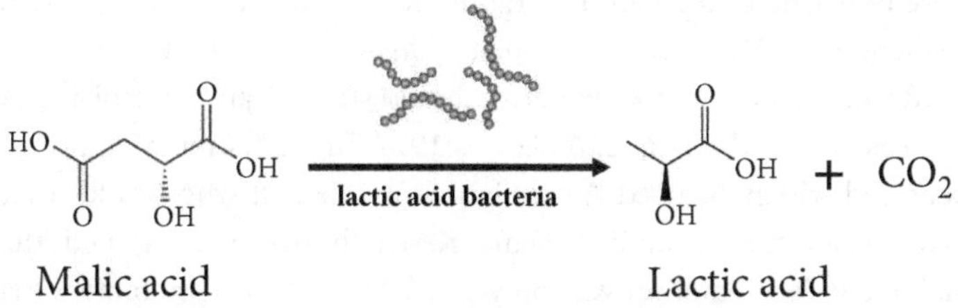

Malic acid → (lactic acid bacteria) → Lactic acid + CO_2

Figure 8.1 Malolactic fermentation. Lactic acid bacteria also produce small amounts of other metabolic byproducts while conducting this transformation, including buttery diacetyl.

familiar with mevushal wines pasteurized for reasons having to do with kosher law. Applying heat under pressure, as is done in "flash" pasteurization, is changing that, though flashing produces its own distinctive flavors that are easy to recognize once you know what you're looking for.

[6] Or "the first stage of the aging process," as it was then thought of, according to Henry Paul in *Science, Vine, and Wine*, 172.

catch-all term for several species that all possess this metabolic capacity. A few have rightfully earned reputations as spoilage organisms for producing un-delicious byproducts. Others, such as *Oenococcus oeni*, have been commercially developed as cultures that can be inoculated in much the same way as commercial strains of wine yeast, *S. cerevisiae*.

MLF has no doubt been happening on its own terms since the dawn of wine, but it's now expressly recognized and deliberately employed. By consuming nutrients that *S. cerevisiae* doesn't, MLF desirably reduces the likelihood of uncontrolled, unwanted microbial growth later on. However, because the process changes a wine's sensory profile, it's not universally desirable. In addition to its distinct flavor, lactic acid is outright less acidic than malic acid. Definitionally, acids release hydrogen ions, H^+, when dissolved in water. An acid's strength is a product of how many hydrogen ions it can release and how easily those ions pop off. Lactic acid has one H^+ to give up, whereas malic acid has two. MLF therefore modestly increases a wine's pH. Some lactic acid bacteria also produce significant quantities of diacetyl, a metabolic byproduct best known for its starring role in butter and artificial butter flavor. All of that works for most reds and rounder white wines—many chardonnays, much pinot gris—but not in crisper rieslings and sauvignon blancs.[7]

In 1939, one of Pasteur's successors, the French enologist Emile Peynaud, made the then-radical suggestion that Bordeaux's wines were best when they had undergone this second, yeast-independent fermentation. He then spent much of his career at the University of Bordeaux advocating that wineries should strategically control it. That was easier said than done. As late as 1977, Bryce Rankine—a prominent and widely beloved Australian wine scientist who would eventually head the Australian Wine Research Institute—argued that malolactic fermentation was only good for microbial stability, often injurious to wine quality, and best avoided when possible, all because it was difficult to control.[8]

---

[7] Diacetyl is much more difficult to detect in red wines than in whites, so if you're thinking that you've somehow been missing loads of buttery red wines: you haven't.

[8] Rankine, "Developments in Malo-Lactic Fermentation." To be fair to Rankine, Australian Wine Research Institute reports from that time suggest that the lactic acid bacteria that most often crossed Rankine's desk happened to be high diacetyl producers that left wine tasting inappropriately buttery.

Controlling MLF has become easier as wine microbiologists continue characterizing the lifestyles of lactic acid bacteria, and thanks to technical improvements in temperature and microbial management. Though they can be inoculated, lactic acid bacteria usually just show up when favorable conditions are cultivated for them: relative warmth (they prefer living above 60°F), not too much sulfur dioxide (they're more SO_2-tolerant than many microbes, but not as tolerant as *S. cerevisiae*), and neither wildly acidic nor wildly alcoholic. The issue remains that these bacteria don't know which wine styles are supposed to undergo MLF and which aren't, so they need to be actively excluded when they aren't wanted. Sterile filtration, antibacterial enzymes such as lysozyme, 🍷 or (and) plenty of sulfur dioxide will do that job. Wineries that forgo those tools need to rely on chilling and other more modest discouragements.

Lactic acid bacteria may now be welcomed, but even they can become spoilage organisms when they show up to work uninvited, like perfectly nice plants that become weeds on a golf course. Except, unlike weeds, weedy microbes can't be found simply by walking around and looking for them. Humans usually can't perceive micro-scale critters with unaided senses unless there are a lot of them, by which point it may be too late, so the precise meaning of "looking" for microbes matters.

Thanks to major changes in how scientists go looking, we now know that lactic acid bacteria sometimes pop up where they're not wanted because they've gone VBNC. VBNC—viable but not culturable—is a suspended state somewhere between life and death, in which cells don't eat or grow, but from which they can be revived under the right conditions. Not all microbes can enter a VBNC state, but several significant wine microbes can. From the microbe's perspective, it's a strategy for surviving stressful circumstances in which their survival would otherwise be marginal, like wine. 🍷 From a human perspective, it can be a real pain, because the presence of VBNC cells is easy to miss until they revive and assert their presence, potentially months after winemakers have declared a ferment safe and sound.

When "looking" for microbes meant spreading a bit of wine (or what have you) on a petri dish and expecting that anything alive would grow, VBNC microbes may as well have been wearing an invisibility cloak. Now that "looking" can mean sequencing DNA to detect the presence of microbial genetic material, metabolically inactive microbes are easier to find, *if* you know that you should be looking in the first place. DNA sequencing is becoming cheaper every year, but it's far from cheap enough to wander around sequencing everything without suspecting that you should.

Controlling More Microbes

DNA sequencing may let wine microbiologists look for microbes in new ways, but they're largely still looking *at* them through old concepts—namely, the Pasteurian paradigm, wherein a good microbe is a dead microbe or, in exceptional circumstances, a well-controlled one. The received wisdom has said that microbes are difficult to control and might become sources of spoilage if uncontrolled; therefore, eradication is the best general rule. The main change across the past few decades is that the number and range of exceptions to that rule are rising.

New, potentially useful microbes are being drawn from grape juice's normal flora that flourish in the first stage of an uninoculated ferment when it isn't loaded up with sulfur dioxide, and before *S. cerevisiae* quashes them by making alcohol. The Pasteurian paradigm considers those initial colonists unwanted spoilers when they're not controlled—hence the sulfur dioxide—but wine microbiologists are now working to tame some of them in the interest of deploying their skill sets strategically. Among a class of yeast species rather unimaginatively (and tellingly) called "non-*Saccharomyces*," for example, *Schizosaccharomyces pombe* eats malic acid and can flatten unwanted sharpness even in wines whose styles don't favor MLF.[9] *Starmerella bacillaris* eats fructose faster than glucose—the opposite of *S. cerevisiae's* preference—and ferments those sugars 🍇 into relatively less ethanol and relatively more glycerol, providing a possible route to lower-alcohol wines. 🍇. *Lachancea thermotolerans* can metabolize sugars directly to lactic acid and produces tropical-fruity thiol compounds along the way, potentially improving acid balance in flabby juice or punching up the aromas of sauvignon blanc. Innocuous fungi such as *Metchnikowia fructicola*

---

[9] *Schizosaccharomyces pombe*—a yeast that divides by fission (splitting in half rather than budding, hence "schizo")—ferments sugars (hence "saccharomyces" or "sugar-loving"), and that was first identified in African millet beer (hence "pombe," meaning "beer" in Swahili).

can be employed as "bio-protectants" by taking up space to reduce the risk of undesirable microbes moving in. 🔊

All of these alt-yeasts have been vetted and developed as commercial cultures. None can carry a fermentation to completion, and *S. cerevisiae* will outgrow them, so the non-*Saccharomyces* is often added first and given a few days to work before a standard fermentation yeast is pitched to finish the job. Microbes such as *S. pombe* whose effects are appreciated in modest doses only can even be added in a form that permits mechanically removing them when their job is done, such as embedding microbial cells in beads, steeping them as you would a tea bag, and then extracting the bag.

This sounds great, but even tea-bagged microbes can't be treated just like any other non-living ingredient. Not only are they living, but, being living, they can interact with one another. The conventional Pasteurian view tends to imagine microbes as independent actors— understandable, if your understanding is predicated on studying individual isolated species. Under real-world conditions, where microbes live in heterogeneous groups known as microbiomes, it's pretty silly. Microbes have social lives. *Brettanomyces*, dreaded among spoilage microbes, provides one example of why we should care.

Microbial Sociality

Brettanomyces bruxellensis, a divisive yeast usually known just as Brett, is infamous for the sensory characteristics associated with its signature metabolites 4-ethylphenol (4-EP) and 4-ethylguaiacol (4-EG): barnyard, wet dog, medicinal, sticking plaster, leather, spice, or smoke. The spicy, smoky end of that spectrum can synergize appealingly with some wines, at least for some drinkers. Enough oenophiles have found Brett's best qualities at least occasionally desirable to warrant investigating whether the difference between stinky and sexy might be "good" and "bad" Brett strains, and whether good ones can be tamed and inoculated in the name of adding complexity.

The answer, at least at present, is a firm no. Brett remains perhaps the least controllable of all well-known wine microbes, so deliberately cultivating it looks more like shooting yourself in the foot than hiring someone new onto the winemaking team. Brett is also able to go VBNC. And even were that not the case, Brett's flavor profile seems to be more a function of environmental conditions than genetic background, so there may be no such thing as a "good" strain.

The metabolites 4-EP and 4-EG are unmistakable signs of Brett; they don't appear in wine otherwise. Their precursors, in contrast, are universal. Brett makes its unique stench by metabolizing hydroxycinnamic acids, a group of phenolic compounds found in many plants and invariably present in wine. However, the sensory outcome of a Brett infection is shaped by the precise kinds and quantities of hydroxycinnamic acids that happen to be around. Other molecules that can feed Brett also matter, including pentoses, five-carbon sugars that *S. cerevisiae* won't touch. 🦠 As with other microbes that live in wine, how *well* Brett lives and how fast it grows depends on available food, alcohol concentration, temperature, and pH—and its relationship to its neighbors.

Like Brett, some strains of lactic acid bacteria also metabolize hydroxycinnamic acids. While they don't produce 4-EP and 4-EG, they have the first half of the metabolic machinery that leads in that direction, so they spit out intermediates one step further down that path. In the company of such bacterial strains, Brett will eat these intermediates and produce even more 4-EP and 4-EG than otherwise (see Figure 8.2). Now that this relationship is appreciated, wine

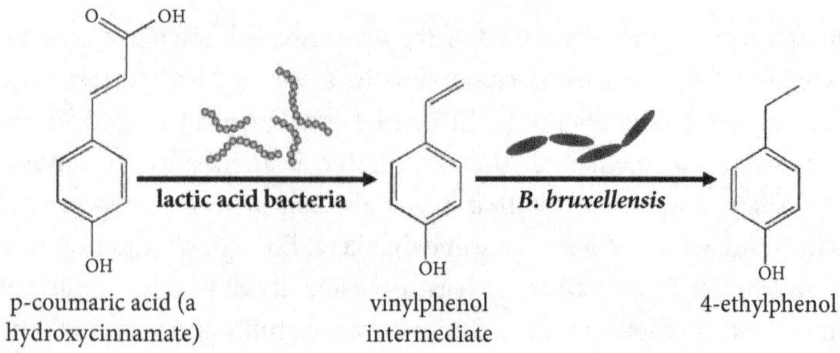

p-coumaric acid (a   vinylphenol   4-ethylphenol
hydroxycinnamate)   intermediate

Figure 8.2 Some lactic acid bacteria can metabolize hydroxycinnamic acids such as p-coumaric acid (ubiquitous in grapes and wine) into an intermediate vinylphenol, which Brett can then metabolize to 4-ethylphenol (or, beginning with a slightly different hydroxycinnamic acid, 4-ethylguaiacol) to produce its signature stench.

microbe suppliers are cultivating lactic acid bacteria for MLF that don't have the offending metabolic capacity.

So-called cross-feeding among microbes isn't unusual, and microbial biotechnologists are engineering similar relationships to force pairs of microbes into stable relationships (see the COMPLEXITY box). This is useful because deliberately co-culturing two microbes is tougher than you might expect. One tends to outgrow and swamp the other. This is certainly true for commercial winemaking yeast, which have generally been commercialized because they excel at domination—ideal for excluding other microbial influences, but unhelpful should a deliberately mixed microbial fermentation be the aim.

COMPLEXITY

One path toward simultaneous complexity and control involves mixing multiple microbes in one ferment. Another is to mix multiple yeasts in the lab by breeding two-species hybrids. But aren't species definitionally unable to mate with each other, you say? That general rule for macroorganisms breaks down for microbes. Really, the whole idea of species breaks down for microbes. Many, including our friend S. cerevisiae, promiscuously exchange bits of genetic material, absorbing DNA from their environments in ways that aren't physically accessible to macrobes like us. They can also occasionally mate in unusual ways.

Yeast don't need to mate to reproduce. Saccharomyces and other "budding" yeasts usually make more of themselves by pinching off one end of a parent cell to form a child cell; mating is primarily a matter of generating genetic diversity. Two cells with different mating types—nothing whatsoever like male and female, even if the comparison is inevitable—respond to extracellular signals (yes, like pheromones), extend toward each other, and form a bridge across which they send DNA. While totally different yeasts won't mate, it turns out that if you let relatively similar Saccharomyces species sit with each other for long enough, a few will get together. The hybrid offspring from such rare matings sometimes have the metabolic capacities of both parents: the

fermentation expertise of *S. cerevisiae*, for example, plus the inclination of *S. kudriazevii* or *S. mikatae* to produce characteristic sets
of flavorful metabolic byproducts. Yeast breeder Jenny Bellon has
shepherded a few such hybrids that the Australian Wine Research
Institute has commercialized.

Yeast hybrids aren't genetically modified organisms. Constructing
them involves plain old breeding techniques—the process is incredibly boring, really—without cutting and pasting genes from one species into another. Genetically modified wine yeasts do exist, though
the vast majority haven't left the lab because they're not approved
for winemaking in countries that subscribe to the Organisation
Internationale de la Vigne et du Vin.[10] In countries that don't—that
would be the United States, Canada, and China—a commercially
available GMO yeast named ML01 can be used to perform alcoholic
and malolactic fermentations simultaneously, having been endowed
by molecular microbial geneticists with the necessary metabolic machinery for the latter from lactic acid bacteria.

Producers who employ ML01 aren't exactly shouting the news
from their rooftops, given how controversial GMOs are in general and how wary the wine industry has been in particular. ML01
raises no safety concerns for consumers. On the contrary, a selling point of having one microbe do the work of two is that ML01
reliably won't produce biogenic amines, human-hormone-like
molecules that some lactic acid bacteria produce and that leave
sensitive humans with a hot flush and a headache. 🍷 Another
major selling point, from the winery's perspective, is that combining alcoholic and malolactic fermentations gets wine out of the
cellar faster. And given that the available evidence indicates that
commercial yeasts aren't overrunning less-domesticated strains
in environments around wineries, ML01 doesn't raise realistic environmental safety concerns, either.

---

[10] The Australian Wine Research Institute even has a yeast strain modified with the
necessary genes to produce raspberry ketone so that it leaves chardonnay smelling like
raspberries—at least according to the researchers who've been the only ones to try it.

If ML01 is controversial, it should be for other reasons, because it changes the bar for efficient winemaking in ways that risk making commercial wines even more standardized than they already are. This isn't much of a risk. Folks who care about cultivating complex microbial communities and complex flavors will do so. Businesses that produce boring wine already have plenty of tools to ensure that their products are unremarkable, if that's what they're going for. Were relying on fewer and fewer microbes to become the norm, however, the wine industry would have two problems. First, homogeneity is vulnerability. Reducing biodiversity increases the risk that a weakness in one creature makes the whole system fall apart. Second, reducing biodiversity reduces long-term cultural richness in the name of short-term corporate profit. Fortunately, for now, the trend is toward expanding the range of microbes employed in winemaking, not contracting it.

Caring About More than Control?

In other words, wine microbiology has addressed one set of problems so well that it's created new ones. Winemaking before the advent of commercial cultures effectively trained yeast and bacteria to work well together, leaving the details of how to arrange stable living arrangements up to the microbes. A lot of twentieth-century (and even a fair bit of twenty-first-century) wine microbiology has been about training a small number of selected microbes to dominate in the interest of trying to exclude everything other than those selected strains. A different, *post*-Pasteurian view of microbiology might instead suggest that the capacity to get along well with others is important after all.

In many, even most cases, microbes aren't *inherently* either beneficial or spoilage organisms. They *become* undesirable when they're allowed to overgrow. Consequently, controlling against spoilage organisms can itself make them into spoilers. (Near-)sterilization eliminates microbial neighbors that might otherwise keep each other in check, leaving space for anything that sneaks past sulfur dioxide and fine-grained filtration to take over. Then, in "clean," fruity wines, microbially derived characteristics stick out at stinky angles where they might blend into wine styles built to include them.

Case in point: yeasts blamed for faults in modern control-oriented production paradigms, such as *Rhodotorula mucilaginosa* and *Pichia anomala*, are par for the course in elegant qvevri wines made using the ancient Georgian technique of smushing grapes into clay jars (*qvevri*) and leaving them to their own devices. *Pichia anomala* contributes to the first stage of many uninoculated ferments, producing a small amount of ethyl acetate that melds with a wine's overall bouquet; unchecked growth, in contrast, overplays that note and leaves wine with eau de nail polish remover. People who enjoy a little barnyard in their Bordeaux (or Barbara, or Napa cab) would even say that Brett isn't always a spoilage microbe. Like *O. oeni* and *P. anomala*, it *becomes* one when what it's doing doesn't jibe with a wine's overall style.

Heather Paxson, an anthropologist at MIT, coined the term "post-Pasteurian" to describe an attitude toward microbial life that follows but deviates from the Pasteurian tradition—microbes are everywhere, and this is an opportunity to live and work well with them, not (just) a challenge to human control.[11] Post-Pasteurians still want to encourage beneficial bugs and exclude pathogens. They want their wine or cheese to taste good and be nutritionally sound and not cause foodborne illness; they want their gut microbiota to support their own good health and the microbiomes in their fields to support the health of their crops. But they don't begin from the position that microbes are dangerous and generally need to be eradicated.

Post-Pasteurians eschew the erstwhile equation of hygiene with killing bacteria, a hangover from Victorian times when germ theory was new. Instead, they embrace the implications of newer microbiological knowledge. Microbes are wildly diverse. Most are just down there doing microbe things without hurting anyone who walks on two legs. Some, like the normal flora that populate mammalian digestive tracts, are cornerstones of life and health.

The "post" part of post-Pasteurian is essential. We're not talking about rejecting germ theory or pasteurized milk. We're talking about

[11]   Paxson, "Post-Pasteurian Cultures."

recognizing that germs are unusual among microbes, and that control isn't necessarily the best approach to avoiding them. Nor, for that matter, is relinquishing control—not that total control over microbial life has ever been possible in the first place, as humanity's persistent inability to eradicate (human or wine) infectious disease makes clear. But it might be that control itself isn't the most useful starting point for microbe-human relationships, at least not for all of them. We could talk about humans and microbes participating in each other's lives. We could talk about microbes as collaborators. We could talk about how humans care for microbes and maybe how microbes care for humans too.[12]

None of that means painting some idiotic sunshine-and-roses view of the more-than-human world in which humans should try to make friends with all microbes everywhere, or in which all microbes want to make friends with us. We don't, nor do they. It means that grouping the wildly diverse collection of critters known as microbes into a single box, let alone a box named "pathogens," is absurd. It means recognizing that as we see more nuance in how the microscopic world works, we need more nuanced ways to work with its inhabitants too.

---

[12] See the work of Emma Frow at Arizona State University on care, for example, Josh Evans's work on interesting microbes at the Technical University of Denmark, or my own recent academic work on participation and collaboration.

9

Alcohol ⊱⋄

Alcohol is a metabolic quirk, for humans and for yeast. It's unquestionably toxic. It's central to everything wine is. Producers have often been looking for more of it. And yet today a major theme across the industry is dialing it back.

All Alcohol Is Not Equal

Chemically, "alcohol" is the name for any organic molecule bearing a hydroxyl or -OH group on one of its base-skeleton carbons. Ethanol is the specific name for the second-simplest member of that group, with just two carbon atoms. Only methanol, with just one carbon, is simpler (see Figure 9.1).[1]

Methanol is much more toxic than ethanol; were some impossibly evil imp to replace the ethanol in a bottle of wine with methanol, one glass could leave you permanently blind. Imps aside, unless you're in the habit of drinking inexpertly distilled moonshine, you needn't worry. *Saccharomyces cerevisiae*, our friendly fermentation yeast, produces smidgens of methanol as an ordinary byproduct of fermentation. While national governments regulate wine methanol content, table wines rarely approach those limits. Only when the products of fermentation are distilled does methanol become concentrated enough to cause trouble, and any reputable commercial process carefully separates the methanol-containing portions of a distillate from the rest.

---

[1] Chemical terminology can feel abstruse, but it's systematic; once you know the basic parts of speech, a compound's name will tell you it's structure; vice versa, if you know the structure, you know its systematic name. "Meth-" is a prefix meaning "one carbon." "Eth-" means "two carbons."

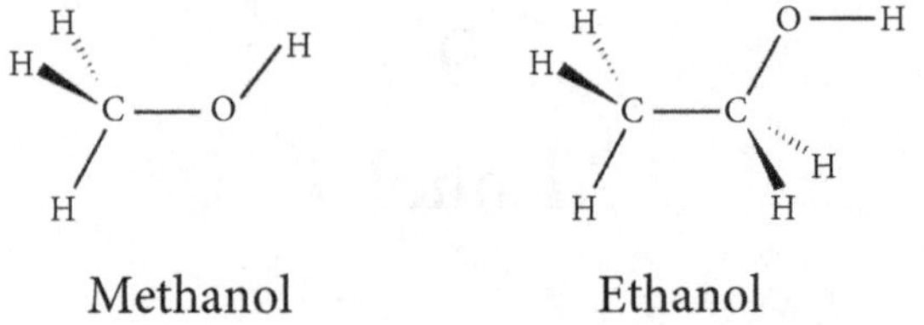

Methanol   Ethanol

Figure 9.1 Methanol, with one carbon, and ethanol, with two carbons.

Larger molecules in the alcohol family, with longer carbon chains, are more relevant for our purposes. These "higher" or "fusel" alcohols, after the German word for low-quality booze (in which they tend to be plentiful), are also byproducts of yeast metabolism and contribute to the varied flavor profiles of different yeast strains. Variations in yeast strains' flavor characteristics can be chalked up, in part, to exactly which fusel alcohols they produce and in what proportions. As the "fusel" moniker suggests, their flavors aren't necessarily pleasant. For instance, 3-(methylthio)-1-propanol smells like boiled potatoes. But some can be floral, and they're often part of a pleasing overall picture in small quantities.

The importance of lower and higher alcohols notwithstanding, we all know that every member of this class isn't equally important. So let's get on with the real business of this chapter—how microbes and humans strategize around the one alcohol that brought us all here.

Making and Tolerating Ethanol, or Not

Biologically, yeast look more like humans than like most other single-celled organisms, so much so that yeast cells are routinely employed as experimental stand-ins for human cells. Our similarities make yeast's metabolic quirks look all the quirkier. Humans and yeast can both extract energy from glucose via aerobic respiration. 🦠 Yeast, however, prefer anaerobic fermentation under most circumstances. This is great from a human perspective, because the primary waste product of fermentation is ethanol. It's weird from a microbial perspective because fermentation is massively less efficient than aerobic respiration, yet yeast use it even when it seems that they're not obliged to.

Metabolic efficiencies and inefficiencies are measured in a cell's production of adenosine triphosphate, or ATP. Cells ingest molecules of glucose or another favorite food and pass them down a series of enzymes that break chemical bonds. Energy released in breaking those bonds is stored by making new bonds, linking an additional phosphate onto adenosine *di*phosphate to make the storage molecule ATP. Later, that high-energy phosphate–phosphate bond can be broken to power work around the cell.

Yeast, were they to use the same oxygen-dependent metabolic strategy that you and I are relying on right now, would obtain something on the order of eighteen molecules of ATP from one molecule of glucose. Instead, when yeast employs alcoholic fermentation, it produces only two molecules of ATP from that same glucose molecule—massively less payoff from the same quantity of food.

Alcoholic fermentation has the salient advantage of not requiring oxygen, but here's the kicker: *S. cerevisiae* uses alcoholic fermentation even when it has access to oxygen. Ferments are sometimes protected from oxygen for stylistic reasons, so that oxygen doesn't destroy flavor-contributing molecules, 🛢 but not to push yeast into making alcohol. On the contrary, as we'll see, yeast need some oxygen to build resilient cell membranes capable of withstanding increasingly alcoholic circumstances. They'll make ethanol whether they have access to oxygen or not.[2]

The trade-off for anaerobic fermentation's energy inefficiency is that ethanol gives *S. cerevisiae* a competitive advantage. Fifty or so major groups of microbes take up residence in a batch of freshly crushed grapes. By the third or fourth day of fermentation, most can no longer be detected; even in the absence of additional antimicrobials such as SO_2, 🛢 other microbes generally can't stand how alcoholic *S. cerevisiae* makes their living quarters. Indeed, *S. cerevisiae* reliably comes to dominate every batch of wine-in-progress, even if it's massively outnumbered

---

[2] Sending ambient air through in-process ferments doesn't cause yeast to ferment any more slowly or to make any less alcohol than it would otherwise, according to a 2021 study in Australian Shiraz. Day et al., "Aeration of *Vitis Vinifera* Shiraz."

when fermentation begins—as it always is, unless a packet of commercial yeast and a human who augments the juice with its contents intervene. 🍇

Ethanol makes most microbial life unlivable by disrupting the usual boundary between a cell's insides and its outsides. As far as we know, every living thing on this planet draws boundaries around itself (or its component parts, in the case of us multicellular types) using membranes built from lipids. Lipids are generally uncharged molecules; their long chains of carbon atoms share electrons equally across their chemical bonds, so no one molecule holds an unequal share of an electron's negative charge. The specific variety of lipid used to build cell walls, however, has a negatively charged phosphate molecule attached to one end—they're phospholipids.[3]

That arrangement comes in handy because a cell's insides are mostly water, H_2O, and its outsides usually are too. In contrast to lipids, water is strongly polar. The oxygen in a molecule of H_2O pulls on electrons much more strongly than hydrogen does, so the oxygen end of each water molecule is a bit negatively charged and the hydrogen end is a bit positively charged. Charged molecules such as water and uncharged molecules such as lipids tend to repel each other; oil and water don't mix. That familiar principle also keeps cell membranes organized. The negatively charged, phosphate-bearing ends of membrane-forming phospholipids orient outward toward water while huddling their uncharged carbon chains together, self-organizing into a wall with a very thin polar edge and a thick layer of uncharged insulation (see Figure 9.2). That arrangement blocks most molecules from moving willy-nilly in and out of a cell.

Ethanol ruins membrane organization by being far less polar than water. The hydroxyl group that makes ethanol an alcohol has a bit of a charge, enough that ethanol and water mix well enough. But ethanol's two-carbon chain is uncharged, and just long enough to slip into and disrupt a cell membrane's lipid insulation.

All this means that in an alcoholic environment, membranes leak. Cells die. And to add injury to injury, proteins also tend to disorganize in alcoholic environments because proteins fold in accord with how

---

[3] Bacteria, fungi, plants, and animals all use phospholipids for membrane-building. One huge feature distinguishing archaea, which look like bacteria but aren't, is that they rely on an entirely different chemical class of lipids to build membranes.

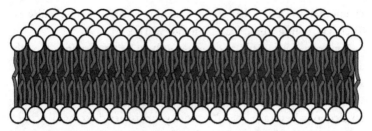

Figure 9.2 A phospholipid bilayer, with charged phosphate heads pointing outward and uncharged, hydrophobic lipids packed inside. Cell membranes are punctuated by channels, sensor proteins, cholesterols that modify the membrane's flexibility, and innumerable other "decorations," but the phospholipid bilayer is the steady background for everything else.

charges along amino acid chains attract and repel. Misfolded proteins maim cells. Every time you rub your hands with an alcohol-based sanitizing gel, you avail yourself of these bacteria-killing physical effects.

You might be asking: if ethanol is toxic, then why can our friendly winemaking *S. cerevisiae* live in it? *S. cerevisiae*, especially strains cultivated for ethanol tolerance, are immune to alcoholic dissolution—up to a point. Yeast respond to the conditions that they themselves create by remodeling their membranes to be more stable and flexible, changing the balance of the specific lipids they employ. Synthesizing these resilient membrane components requires some oxygen; therefore, yeast can't reproduce in alcoholic environments where none at all is available. This is why even wines made in a "reductive" style, deliberately excluding oxygen along the way, 🛢 need opportunities to incorporate oxygen before fermentation begins.

Cell wall adaptations aren't yeast's only alcohol-aware move. They also take other actions, though microbial geneticists and physiologists remain fuzzy on the details. That's kind of odd, given just how important yeast's alcohol resistance is, but perhaps it reflects sensible research priorities, since alcoholic fermentation has worked reliably for millennia, and as the adage goes, science has yet to cure cancer. Then again, in 2016, Yoshinori Ohsumi won the Nobel Prize in Physiology or Medicine for research that began as an investigation of yeast physiology and ended with an explanation of why some cancers resist chemotherapy and a better understanding of Alzheimer's disease. There's no telling where "basic" biology will go.

Your own cells would also react poorly to being suspended in ethanol. Indeed, in a procedure called alcohol septal ablation, cardiologists apply high-proof ethanol directly to heart muscle to kill tissue and alleviate health disorders caused by cardiac overgrowth. Happily, drinking an alcoholic beverage doesn't expose your tissues directly to alcohol. Saliva dilutes beverages straightaway. Your digestive tract is also coated in a layer of protective mucus.

We could take all day on the finer points of alcohol's effects on humans, but I want to return to yeast, so I'm going to move more quickly through this part. Ethanol diffuses from the gut into the veins that line your digestive tract, which run to the liver rather than directly back to the heart the way that most veins do. Enzymes in the liver break ethanol down, first into acetaldehyde and then into acetic acid. Acetaldehyde is itself toxic, and acetaldehyde buildup in the bloodstream seems to be responsible for many short-term effects of overconsumption: flushing, nausea and vomiting, headache, and so on. Your personal ethanol metabolism rate (affected by how much you've drunk, what you've eaten alongside, other drugs you're taking, and your individual physiology) also seems to be a factor, since ethanol left undigested in the bloodstream can reach the brain and lead to toxic metabolites there directly too.

Digesting alcohol is primarily the liver's task, but it can happen in most cells—and will, if you've drunk too much. Everywhere it happens, the process generates unstable, highly reactive oxygen radicals. These oxidants can cause all manner of indiscriminate damage, putting widespread stress on cells. Chronic exposure to this oxidative stress is behind the systemic health effects of living with alcoholism, including liver damage and neurological dysfunction.

Competition, Obviously

I've made it seem as though competition is the obvious solution to the strange business of alcoholic fermentation, but that isn't quite the full story. Competition is a fine explanation for *how* ethanol production leads yeast to dominate fermentations even when initially outnumbered. But competition isn't an explanation for *why* yeast produces ethanol. To make that logical leap, we'd need to imagine that yeast had thought about its options in life, realized that it was interested in setting up dominion over grape juice and that ethanol would help, and then acquired the metabolic machinery to produce it. This is obviously not a sensible biological explanation.

Competition is always worth a second look in biology because competition is *the* default explanation for how living things interact, from how two species of squirrel compete for one environmental niche to how two molecules compete to bind to an enzyme.[4] Because it's the default, the way the fabric of biological knowledge tends to suggest things *should* work, saying "X is about competition" is an easy gloss for "we don't understand much about how X works." And in yeast's case, there's plenty that's not understood.

For example, yeast will only stop producing alcohol when the amount of sugar available for them to eat is very low. In another instance of unexpected research links between cancer and yeast, that phenomenon was first observed in cancerous tumors by a London-based biochemist named Herbert Grace Crabtree in the 1920s.[5] The Crabtree effect functions as a predictable exam question for every wine microbiology student. Its function for yeast is less obvious.

One possibility comes from the relative length of the metabolic pathways in question. Aerobic respiration requires many enzymes. Anaerobic fermentation requires far fewer, making the former a more expensive investment in terms of metabolic infrastructure. Space constraints are also an issue, since these enzymes need to be associated with a membrane to stay close to each other so that molecules can be passed down the enzymatic chain. A cell only has so much membrane to go around.

A second consideration is that fermentation is fast. Using it allows yeast to rapidly consume microbially common goods while squirreling away private resources for later.[6] Every microorganism you'll

---

[4] Competition is also sometimes vague or misleading as a biological term, because those squirrels or molecules may "be in competition" with each other without ever directly interacting if they're aiming to use the same resources.

[5] Crabtree, "Carbohydrate Metabolism."

[6] Dashko et al., "When, Why, and How"; Piškur et al., "How Did *Saccharomyces* Evolve."

encounter in grape juice can obtain energy from sugar. Few can use ethanol in the same way, but *S. cerevisiae* can. Converting sugar to ethanol lets yeast both starve out and poison other microbes while storing a food source for later.

Alcohol seems to be all about competition, but exactly what a "competitive advantage" means is context-dependent. Only very rarely does "competition" entail two creatures actively battling it out, yet the idea that living things have to compete to survive in a hostile world is easy to read into the language we use.[7] Relatedly, biology's foundational principles were built by wealthy European men whose cultures reified the domineering individual and whose countries were pretty well always at war. Being satisfied with "competition" as an explanation for *why* biology works the way it does makes it all too easy to forget context, biological and cultural, and to accept lazy logic.

In high school, I was taught the "central dogma": DNA makes RNA makes protein. Molecular geneticists now know that while it's not entirely wrong, it's massively incomplete. DNA doesn't *make* anything; DNA is used as a template by proteins and RNA molecules. RNA has innumerable intracellular roles beyond acting as DNA's messenger. The old idea that one segment of DNA produces one protein—"one gene, one enzyme"—is wrong, because the same bit of DNA can have multiple functions depending on its interactions with protein and RNA in concert with whatever is going on in the cell. In short, molecules interact not in hierarchical chains of command, but in complex networks of informative relationships. Such networks are easier to see with newer techniques that allow biologists to study systems instead of one molecule at a time. They're also part of a bigger move away from imagining one individual in control of its existence, fighting for survival, toward seeing how virtually everything in life happens through complex interactions.

That core idea, that life is a function of myriad interacting factors, is true for microbial societies too. 🐚 Microbial neighbors don't just compete with each other for resources; they also cooperate, sharing food or secreting defenses that protect whole neighborhoods. Context

---

[7] Keller, "Language and Ideology."

matters; one day's friend may be another day's enemy. Two microbial neighbors may simply be indifferent to each other.

Humans make sense of microbial relations through concepts such as competition and cooperation, initially developed for thinking about fellow humans. That's understandable. Those are the tools we have. But in the ongoing scientific quest to build more detailed, nuanced, useful knowledge about how the world works, we need to keep asking: are we assuming that other creatures live and work the way that one idiosyncratic subset of humanity does?

Nolo?

If producing alcohol gives *S. cerevisiae* an edge over other microbes, the opposite is becoming true for alcoholic beverage producers. The wine industry is highly competitive, with hordes of brands battling it out for consumers' attention. But maybe competition isn't that simple in this ecosystem, either. Producers aren't all trying to claim the same consumers, and some of their moves—like the newish shift toward lower- and no-alcohol offerings—may be about building new niches, not ousting a competitor from an existing one.

The market for no- and low- or "nolo" wine-like products (more on why I need to say "wine-like" later) is booming, after a few decades of steadily rising numbers next to the letters "ABV" on labels. That familiar acronym stands for "alcohol by volume," the standard industry unit for reporting how much ethanol a liquid contains. (The ethanol fraction could just as well be *weighed* and compared to the mass of the whole, instead of comparing the volume of alcohol to the volume of water; that's just not how it's typically done.) Plenty of ABVs are the product of alcohol adjustments, thanks to a spectrum of technologies for reducing or removing ethanol and a spectrum of motives for applying them.

Sales of no- and low-alcohol wines have been rising and, as of 2020, constitute about 3 percent of the global wine market. If you find yourself in that market, you may have been clobbered one too many times by a bombastic 16 percent cabernet sauvignon. Or maybe you're trying to limit your alcohol consumption without curbing your wine habit.

Or maybe you're watching calories and mindful that higher-alcohol wines contain more of them.

To be clear from the start, alcohol reduction doesn't just apply to "nolo" products. Plenty of ordinary table wines—including upward of 20 percent of Californian specimens—have taken a trip through post-fermentation alcohol mitigation. Even more invoke ethanol-reducing strategies in earlier stages of winemaking. Either way, we're seeing some combination of a response to warmer growing seasons and a response to waning appreciation for alcohol bombs—though interest in wines of drinkable strengths is in tension with the seemingly eternal fad for big, ripe flavors.

Ripeness

In Italy's Veneto region, blooming happened thirteen to nineteen days earlier in 2009 than it did in 1964.[8] Growers might just pick earlier, and they have been; globally, harvest dates have advanced. But earlier harvesting isn't enough to keep wine styles constant. Daytime temperatures have also been rising, causing grapes to accumulate sugars faster. Vines regulate sugar accumulation and the development of grape color and flavor development separately, so prematurely sweet grapes generally aren't developing other ripeness benchmarks at the same accelerated rate. Consequently, producers who choose harvest dates on the basis of flavor will contend with more sugar and more alcohol. Harvesting when sugars align with a desirable final alcohol concentration is liable to leave green flavors instead. The upshot, often, is higher ABVs on labels.

Older wine research routinely worked with "model" red wines in the neighborhood of 12 percent ethanol. Today, industry relevance means raising those reference figures to something more like 14 percent. But rising temperatures aren't acting alone. To some extent, alcohol content is traveling upward in response to the so-called international style of ripe, fruity, full-bodied wines, which under many circumstances can only be achieved through more alcohol. A 2012 analysis found that ABVs were escalating much faster than warming climates could

---

[8] Tomasi et al., "Grapevine Phenology."

account for, more or less worldwide. The same study found that producers were leveraging the legally acceptable margin of error between the ABV on the label and the precise ethanol contents of the wine to match perceived expectations about how potent particular styles of wine *should* be, understating alcohol content for big, ripe reds and (to a lesser degree) overstating it for lithe whites.[9]

In the grand scheme of wine research, the new push for alcohol-reduction strategies is ironic, because past research has generally worked to *increase* alcohol. Or, we could call it a matter of restoring balance to the universe, now that preferences for bigness and the research that's enabled so much of it are beginning to top out and turn the other way.

Some ripeness-seeking research has been in service of the challenges of cool climates. / In upstate New York's Finger Lakes region, growing European *Vitis vinifera* varieties was thought impossible before the 1950s because vines would freeze in winter; now, thanks to viticultural developments aided by Cornell and other northerly universities, the region is a world leader in Riesling and even grows some respectable Pinot Noir and Cabernet Franc.[10] Elsewhere, revised vineyard techniques have served a shift toward fruitier wines that drink well sans cellaring (see the TASTE box). But now, cooler climes such as Germany are dealing with higher alcohols than their signature styles were built to accommodate, while fruit-forward wines from warmer spots can be undrinkably overripe. To some extent, the lower-alcohol trend is a course correction after viticultural research and the climate crisis colluded to succeed too well at maximizing sugar accumulation.

The other half of that story is that reducing alcohol has become desirable for multiple reasons. Some alcohol adjusters are motivated by a better-for-you vibe on the shelf. Many want to harvest when grapes have desirable flavors and then engineer a wine's ethanol content for sensory balance later, so that they don't have to fret about alcohol while deciding when to pick. Ethanol sometimes manifests as an overtly alcoholic edge, but it can also dampen fruitiness or mask varietal character,

---

[9] Alston et al., "Splendide Mendax."
[10] Winter hardiness remains an issue, and universities in Minnesota, Michigan, Wisconsin, and New York, among others, have active research programs in developing wine grapes that don't mind the cold.

TASTE

Anyone who needs to taste dozens of wines in one sitting—including judges at wine shows—will find that wines with the biggest, boldest flavors stand out, simply by being best equipped to rise above the palate fatigue that inevitably sets in under such circumstances. When wines are marketed on the basis of how many gold medals they've been awarded, consumers will come to equate quality with wines that show well, even if show-worthy flamboyance doesn't equate to drinkability over dinner. Shifts in drinking habits have also favored bigness, since the wine that drinks well over dinner may not be the same that appeals as a solo glass during cocktail hour. In addition, wines that sell on being fresh and fruity are best drunk young, whereas wines that prioritize structure are more likely to benefit from cellar time, and few folks are cellaring their wines; 90 percent of wines sold for off-premise consumption in the United States are drunk within two weeks of purchase.

so adjustments are sometimes a matter of achieving sensory balance. Sometimes alcohol reduction is a cost-saving move, though this motive has recently become less motivating in the United States. Wine is taxed on the basis of how much alcohol it contains. The jump between brackets is enough that an investment in a few tenths of a percent of alcohol reduction can more than pay for itself. Before December 2020, tax rates jumped by about 50 percent for wines over 14 percent ABV. Now that threshold is 16 percent, so fewer winemakers will need to fight with their wines to pull them under that bar.

Reducing Alcohols

Winemakers who are merely looking to reduce alcohol rather than to remove it entirely have four possible points of intervention:

- In the vineyard: manipulating growing parameters such as pruning and irrigation can reduce sugar accumulation while maintaining flavor and color development, at least to some extent.

- In the winery, before fermentation: adding water to juice is a direct way to dilute sugars.
- In the winery, during fermentation: yeast strains are being developed to convert sugar to ethanol less efficiently.
- In the winery (or a processing facility), after fermentation: ethanol can be physically removed from finished wines.

In the vineyard: Various permutations of alternative vine trellising, strategic leaf removal, yield adjustments, irrigation programs, and harvesting strategies can lead vines to photosynthesize less and so produce fewer carbohydrates to stash in fruit. (So can shade nets, which temper sunlight and vines' capacity to store sugars, though growers using them for hail protection in cloudy spots might wish otherwise.) 🐿 Generally speaking, less leaf area per grape yields less sugar per grape. In practice, that can entail ripping healthy leaves off vines. The challenge, with this and all viticultural strategies, is achieving fewer sugars without unduly affecting flavor. Plants are complex and responsive; as a rule, you can't surgically intervene to alter one and only one plant behavior without affecting others. Even though sugar accumulation is under different physiological controls than flavor and color development are, measures to limit the former often entail changing the latter in ways that may entail more compensatory adjustments in the winery to maintain a desired style.

Before fermentation: If starving grapevines seems violent to some people, adding water to crushed grapes may seem to violate the integrity of a wine to others. When vintners can add water, and how much, is regulated in most wine regions against deceitful producers seeking to fraudulently increase volumes. Yet modest water additions within legal limits can reduce ABV by as much as 1 or 2 percent. Fantastically enough, trained sensory panels tend to report that the results are *more* pleasant and not just weak-sauce relatives of their unadjusted versions. ⌇ Superripe grapes can also be diluted with underripe ones, though unlike water, green grapes bring their own flavors that aren't necessarily wanted.

During fermentation: Having selected winemaking yeast for their ability to metabolize sugar to ethanol, to some extent we can select strains that are less efficient at their job. 🐿 Taking a yeast's-eye view, that looks like a request to shunt more glucose through minor metabolic side paths that produce byproducts other than ethanol, such as glycerol or carbohydrates. Those byproducts will contribute to wine character in ways that will need accounting for. The bigger challenge on this front is that *S. cerevisiae* really likes making ethanol. Hybridizing it with a less enthusiastic but close-cousin species is one route around that preference. 🦠 Another is to give some non-*Saccharomyces* yeast a head start on eating sugar before adding a workhorse *S. cerevisiae* strain to finish up—though the sensory contributions of that other yeast's metabolic byproducts then need to be contended with too.

After fermentation: Ideally, the final set of options—physically removing ethanol after fermentation—would precisely adjust alcohol without altering anything else. Wine engineers have yet to achieve that ideal, however, because wine itself isn't an "ideal" solution; its components aren't distributed uniformly in water. Wine contains a relatively small but highly significant proportion of phenolics, including astringent tannins and colorful anthocyanins, that tend to group into multi-molecule megastructures. They self-organize as a function of their respective ratios and in response to the overall composition of the "wine matrix," especially pH and ethanol, both of which affect how the charged portions of phenolic molecules interact. The same principle applies to some other wine components. Even ethanol forms little groups on account of being less charged than water. All of this means that any effort to physically remove alcohol and only alcohol alters the physical structure of the wine.

Winemakers may avail themselves of multiple more or less selective ethanol removal strategies, but reverse osmosis, RO, is the most popular. Other options require massive and massively expensive equipment that isn't easy to truck around, literally, as a mobile service on the back of a truck (a necessity for serving smaller wineries). RO is also useful for removing and adjusting far more than just ethanol. 🍷 And at their most basic, RO systems are simple enough that you can buy domestic versions to remove dissolved solids from your water at home.

RO employs the same principle that your kidneys use to recover water from the waste liquid that you'll eventually excrete as urine, and that moves oxygen into barrels.[11] 🛢 Wine is flowed along one side of a

---

[11] "Diffusion" and "osmosis" both describe molecular movement from higher to lower concentration. "Osmosis" is the more specific word for diffusion that occurs across a semi-permeable membrane.

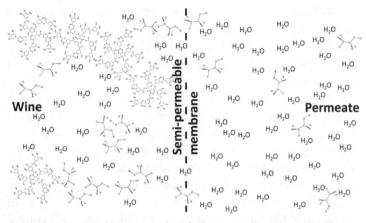

Figure 9.3 Reverse osmosis involves flowing wine down one side of a membrane with pores large enough to permit water and ethanol to pass through but not larger molecules. Ethanol molecules will flow down its concentration gradient, crossing the membrane from the wine into the water.

semi-permeable membrane with water on the other (see Figure 9.3), establishing a concentration gradient between the two. Molecules dissolved in the wine, such as ethanol, will diffuse across the membrane into the water—from the side on which they're more concentrated to where they're more dilute—so long as they're small enough to pass through the membrane's pores.[12] The trick is selecting a membrane with pores big enough to pass ethanol (or whatever else you might want to remove) without being so big as to permit the movement of the desirable dissolved solids that make wine taste like wine.

The good news for RO is that ethanol is small and many other wine molecules are large. The bad news is that no membrane-based technology currently in existence is smart enough to remove ethanol and nothing else. At present, the best way to deal with that limitation is often to process the heck out of a fraction of a batch of wine, removing

---

[12] Both size and charge matter to filterability, since a molecule that's clinging tightly to others or that's surrounded by a shell of water molecules has an *effective* size much larger than the molecule itself.

essentially all of its alcohol without worrying too much about its overall flavor, and then blend that fraction back into the whole batch until the desired ABV has been "dialed in." Advanced chemical separation methods are also making some headway on recovering aroma compounds lost during the process.

Brute-force post-fermentation dealcoholization is the only option for making no-alcohol "wine." If you've ever poured yourself a glass of one of these at a barbecue—the best place to try one, since "wet" and "cold" go a long way under the right circumstances—you've had a firsthand encounter with the problems they incur. They don't taste very good. That's not why I relegated "wine" to scare quotes, though. Per US regulations, anything sold as "wine" must contain at least 7 percent and no more than 24 percent alcohol by volume. The supranational Organisation Internationale de la Vigne et du Vin sets the floor at 8.5 percent, with exceptions for a few traditionally low-alcohol wines. Otherwise, wine-like beverages resulting from major alcohol reduction have to be specially labeled. In the United States, "dealcoholized wine" or "alcohol-removed wine" indicates beverages with less than 0.5 percent ABV; "alcohol-free" is used only for those with no detectable ethanol at all. Similar terms are used elsewhere.

Even *if* ethanol and only ethanol could be surgically removed (which it can't be), the quality of dealcoholized wine would still change because alcohol does more than affect microbial and human physiology. Ethanol tastes slightly sweet, adds viscosity so that wine doesn't feel like water in your mouth, and augments perceptions of other flavors—this last point being the rationale for vodka-tomato pasta sauce. Producers sometimes try to compensate for these absences and everything else lost to dealcoholizing by adding sugar or other fruit flavors—utterly disallowed in anything called "wine," but fine in products called something else. 🍇

Thankfully, modest alcohol reduction in service of sensory balance or drinkability takes us to a very different place than total removal; unless you have superpowers, you'll never know when the wine in your glass is the product of subtle alcohol engineering. In between lies a gray zone of major but not total dealcoholizing, usually in the name of supporting a healthy lifestyle. Research about how best to produce such low-alcohol options often begins with the goal of replicating the ripe qualities of a higher-alcohol wine at a lower ABV. Yet some studies find that drinkers are quite happy with wines with greener flavors or less intensity.[13] Some wines also hold up better to dilution.

It's no accident that New Zealand sauvignon blanc leads the low-alcohol category. New Zealand has made a particular effort to promote "lifestyle" wines, but it can take that tack because Marlborough sauvignon blancs are often so maxed out on varietal distinctiveness that they can be diluted and still meet consumer expectations for a sufficiently "savvy" experience.[14] On top of that, quaffers of "crisp white wine," especially those who want to quaff one that comes in around 7 percent ABV, don't necessarily need much character to be satisfied. I've tried reduced-alcohol Marlborough sauv blancs. They're fine. Dilution is unlikely to change how you feel about this hit-you-over-the-head style.[15]

As someone who likes wine and a clear head, I like the idea of lower-alcohol wine, though I prioritize texture in what I drink and post-fermentation alcohol adjustment often doesn't. Texture aside, the trouble with engineering alcohols—or any other aspect of a wine's character, for that matter—is that engineering makes it too easy to target top-selling styles. If global wine is ruled by the loudest voices, the ones that speak in gold medals and A+ points ratings for wines built to shout above the crowds, then producers are incentivized to play to the common denominator even if there's no direct competitive rationale for doing so. The result is a wash of skillfully engineered, similarly styled, perfectly drinkable, entirely boring crisp white wines. I'm not afraid that big-brand purveyors of engineered wines are going to take away my pét-nat.[16] All the same, I'm sorry when I see that the options presented to the wine-drinking masses teach them that fruity and simple is best when wine ecosystems afford so many other, quirkier options to explore.

---

[13] Reviewed in Varela et al., "Strategies for Reducing Alcohol."

[14] In New Zealand, land of shorthand slang, the country's signature wine is abbreviated "savvy."

[15] Marlborough sauvignon blanc's distinctive character didn't just happen on its own. It's the deliberate result of a national research priority to first identify exactly which compounds are responsible for that distinctiveness and then develop viticultural and winemaking techniques to turn them up to eleven.

[16] Pet-nat, short for *pétillant naturel* or "naturally sparkling," is a category of sparkling wines made by bottling wines for sale with a small amount of residual sugar, and unfiltered, so that yeast finish fermentation in the same bottle you open on your table. They're usually naturally low in alcohol.

10

Sulfur SO₂

You've probably experienced sulfur dioxide by way of back labels. The disclaimer "contains sulfites" contributes to the small print on virtually every bottle, box, or can of wine you can pluck from a shelf. Sulfites are added to most wines as the industry's go-to antimicrobial and antioxidant. Even if they aren't, most yeast strains produce enough sulfites as fermentation byproducts to require labeling anyway. Either way, unless you're among a fraction of a percent of sensitive people, you probably don't sense those sulfites when you drink. Yet you may well have heard the concern, or had the concern, that they're harmful—to humans, the wine, or both—and that the practice of adding them should be curtailed. Meanwhile, wine scientists have been saying the opposite, arguing that sulfites are perfectly safe for nearly everyone and invaluable for quality winemaking, setting up a pro-sulfites versus anti-science dichotomy.

I'm not here to plump for either side because I think that the idea of setting up two sides is silly and misguided. As with most contentious subjects, sulfur lives in the land of "it depends," neither villainous nor a non-issue. Setting up dichotomies is also an easy way to first stereotype and then demonize "the other side," which happens regrettably often.

Most people have more nuanced opinions about most things than either yes or no, in favor or opposed. Wool hats? Yes in winter, no in summer. Black tea? Absolutely if it's hot, no thanks if it's cold. Animal testing? Maybe unavoidable for some medical purposes, but unjustified for cosmetics. Genetic modification? You may support engineering microbes to make cheaper pharmaceuticals without being on the side of Monsanto when they sue farmers who save patented GMO seed from one year to the next. That's the premise I'm starting with. Where I'm going is toward how less sulfur is being used in ways that aren't "against the science," but that are enabled by more nuanced wine microbiology, alongside wider-ranging ideas about what constitutes good wine.

Before we go any further, a note on terminology. In a public health and safety context, we're usually talking about "sulfite," SO_3^{2-}, because this is the sulfur-containing compound typically added to food and drink, often as a stable salt in conjunction with a positively charged ion such as potassium (K^+) or calcium (Ca^{2+}). (Subscripts tell you how many of an atom are in a molecule; superscripts give you the ionization state of that atom.) Food and beverage laws require labeling the presence of sulfites in comestibles. In a wine chemistry or microbiology context, in contrast, the conversation is usually about sulfur dioxide or SO_2, because SO_2 is the actively antimicrobial form of sulfur in wine. Sometimes "sulfur dioxide" is abbreviated to "sulfur," even though sulfur is also present in wine in other forms. If all of that seems confusing, remember that sulfites, sulfur dioxide, and sulfur tend to be interchangeable outside a technical wine chemistry context—and, if it's any consolation, that a lot about sulfur is confusing.

What's the Issue with SO_2?

Sulfur dioxide, SO_2, became an issue long before the rise of contemporary wellness movements, for good reason. Large sulfur additions leave an unpleasant prickly sensation behind, and past winemakers seem to have sometimes overdone it. Romans warded off spoilage with "vapor of sulfur" by burning sulfurous candles inside wine barrels, according to Pliny the Elder's natural history of the known world.[1] No word from Pliny on whether the practice sometimes left people sneezy. But in 1487, a Bavarian royal decree approved adding up to one-half ounce of pure sulfur per fermentation tun but no more, and other German Diets followed with their own edicts against excessive use in the decade or so thereafter.[2] When candles were replaced by mechanically pumping sulfur gas through juice in the late nineteenth century, warmer-climate wineries routinely used enough to leave a bad taste,

---

[1] Plinius, *Naturalis historia* XIV, 129; Paparazzo, "Philosophy and Science."
[2] Lück, "Sulfur Dioxide."

stun *Saccharomyces cerevisiae*, and delay the onset of fermentation, even though wine yeasts are remarkably sulfur-tolerant and easily withstand the quantities used today.[3]

Candles are an imprecise dosing strategy, though still sometimes used to fumigate barrels against spoilage microbes that can be all too happy to hole up in the wood's pores. Gas pumps are no better. Contemporary winemakers have generally given up on vapor-based strategies in favor of adding liquid sulfur dioxide solutions or powdered or crystalline salts such as potassium metabisulfite. Both yield the same chemical effects. Potassium metabisulfite dissociates into its component ions in juice or wine. The potassium ions add to the potassium that comes from grapes themselves. The sulfite ions have several destinies. Some bind to other molecules such as acetaldehyde or sugars. Some react with oxygen to generate sulfur dioxide. Some of the sulfur dioxide then reacts with water to re-form sulfite ions. As we'll see, the remaining "free" sulfur dioxide fraction is the active antimicrobial and antioxidant purpose of this whole exercise. But since sulfur switches among these forms in proportions driven by the specific constitution of a juice or wine, establishing how big a dose is necessary to achieve a desired effect is tricky.

No one who's suffered a faceful of potassium metabisulfite would call the stuff harmless. It's sometimes sold as a white powder, a bit like confectioner's sugar, and with the same tendency to create a cloud if not handled gently. If you've made an American buttercream frosting with an electric mixer, you've probably experienced the moderately unpleasant sensation of inhaling an airborne powder. Electric mixers don't come into play, but handling powdered potassium metabisulfite without respiratory protection delivers a similar effect, only worse. Sulfite stimulates the trigeminal nerve, a bit of anatomical equipment responsible for tingly, sharp, sneezy, almost-but-not-quite painful chemical-induced sensations from your face. The sensation can stick with you for hours.

Inhaling the stuff is a non-issue for people on the bottle-opening, wine-enjoying end of the situation. So what *is* the issue on the wine-enjoying

---

[3] Simpson, *Creating Wine.*

end? For some people with asthma, consuming even small quantities of sulfites will induce airway muscle spasms and difficulty breathing.

Allergy and asthma researchers began properly chasing up anecdotal accounts from suffering sippers in the 1980s. Altogether, they concluded that about 5 percent of people with asthma will experience symptoms within about an hour of imbibing a wineglass's worth of SO_2.[4] The US Centers for Disease Control and Prevention report that about 8 percent of American adults have asthma, which makes the sulfite-sensitive fraction of the population about 0.4 percent, though the exact numbers involved here are a matter of some debate.[5] Some studies have suggested that the proportion of sensitive adults may be as high as 1 percent, including some who don't have asthma. Others estimate that proportion at as little as 0.04 percent.[6] We can blame variation in experimental design for the discrepancy, since the threshold varies for how much SO_2 someone needs to consume before they show symptoms.

Contrary to some popular beliefs, SO_2 hasn't been experimentally associated with wine headaches or with other kinds of intolerances. Nevertheless, sulfites aren't harmless for people who aren't sensitive; in large quantities, they've been associated with gastrointestinal disorders and some cancers. The World Health Organization recommends that people consume no more than 0.7 milligrams of SO_2 per day for every kilogram they weigh. In Australia, the relatively high sulfite additions permitted in processed meats and dried fruit make those foods the largest contribution to dietary sulfite intake for many omnivores. In Brazil, fruit juice is the issue. And in traditional wine-drinking countries, wine will be a major contributor for many people. The thing is, the WHO's guidelines are predicated on experiments in which mice and rats are fed high doses of SO_2, more than a human would consume without taking a spoon to a packet of potassium metabisulfite. We have no direct evidence that sulfites damage humans who consume smaller quantities under ordinary dietary circumstances—though such evidence isn't exactly easy to generate without subjecting people to unethical risks.

[4] Gunnison and Jacobsen, "Sulfite Hypersensitivity."
[5] Estimating the global prevalence of asthma is difficult because the disease is chronically underdiagnosed in many countries, but the World Health Organization estimated that about 262 million people were affected in 2019.
[6] Grotheer, Marshall, and Simonne, "Sulfites: Separating Fact from Fiction."

All of this is to say that it's not fair to declare sulfites innocuous. It *is* fair to say that the significant fraction of wine drinkers allergic to the very mention of sulfites is totally out of proportion to the number who might be realistically indisposed by consuming them. And yet that attitude didn't come from nowhere. Sulfites have a PR problem for a reason.

The US Food and Drug Administration approved sulfites as "generally recognized as safe" for food use in 1958, on their very first list of acceptable food additives. In the following years, sulfites were widely used as a preservative, in dried fruit and cured meats as has been done for centuries, and in an endless array of modern processed foods. Restaurants also started spraying sulfites on salad bars and adding them to fruit and potato salads and such to hinder browning. Sadly, lack of regulation, inadequate training among hospitality workers, and absurd recommendations for how these sprays might be used led to situations in which salad eaters were being exposed to quantities of SO_2 that are inconceivable in contemporary food production.[7] (Processed foods were far less of an issue because factories follow strict recipes, whereas restaurant employees were in a position to spray down their offerings haphazardly.) In 1982, the *Journal of the American Medical Association* published a report on a teenage American girl who had been hospitalized, twice, after eating in restaurants on an Italian holiday. Four people died from severe salad-related reactions in the United States in 1983 and 1984 alone.[8] Researchers started searching for less severe sulfur-related incidents and found them, including adverse reactions to wine. The FDA finally responded by outlawing the all-too-easy-to-abuse salad sprays in 1986 and by requiring that other sulfite-containing foods and drinks be labeled as such.

As of 1993, the legal upper limit for SO_2 in wine sold in the United States has been 350 milligrams per liter. The same number applies in Hong Kong, Japan, and parts of Canada (Ontario and Quebec have their own rules). In the EU, it's 150 mg/L for dry whites, 200 mg/L for dry reds, and more for sweeter wines. In Australia and New Zealand,

---

[7] Martin, Nordlee, and Taylor, "Sulfite Residues in Restaurant Salads."

[8] Sun, "Salad." Sulfur had been used in the food industry before the 1980s, but the popularity of restaurant salad bars and sprays seems to have brought the latent issue to a head.

it's 250 mg/L for dry wines and 300 mg/L for sweeter ones. Barring the occasional rookie error, the idea of a decently made commercial wine in the twenty-first century exceeding the US legal limit is laughable—unless you see sulfur as being no laughing matter under any circumstance. That would be understandable if you're sensitive. Other folks might still be suffering an anxiety hangover from the 1980s.

Unfortunately, mandatory "contains sulfites" labels—essential information for a small number of people—may give others the idea that sulfites are bad, following the logic that special labels are only going to happen for things you shouldn't be eating. In addition, in the transitional period following new labeling requirements, European wines from earlier vintages didn't carry the same warnings that were appearing on American wines, which made it look as though Americans were adding extra preservatives. They weren't. They still aren't. Winemakers use sulfur the world over. That label isn't there for you—unless it is, in which case you almost surely know who you are.

So, Why Are Winemakers Using It Anyway?

All the same, one might wonder why winemakers still use the stuff. The thing is that sulfur is a Swiss army knife for winemaking, right down to having some functions that are confusing or poorly defined. For many winemakers, its functions, well-described and otherwise, are built into the way wine is made. For the same reason, the move to use less should be welcome, even if only in the interest of exploring stylistic possibilities.

SO_2's sustained popularity is warranted by oxygen and microbes. Its activity against these two sometimes-threats is separate, so let's deal with oxygen first. Oxygen is toxic. The history of life on earth since the emergence of photosynthesis is a story of creatures dealing with the aftermath of oxygen entering earth's atmosphere.[9] Even creatures

---

[9] Photosynthetic microbes began producing enough oxygen to alter earth's atmosphere about 2.4 billion years ago, in what some scientists call the "Great Oxygenation Event" (Hohmann-Marriott and Blankenship, "Evolution of Photosynthesis"). Land-dwelling plants as we understand them today are a comparatively recent development, emerging only about 500 million years ago.

that have come to depend on the stuff require substantial antioxidant protections against its damaging effects.

Atmospheric oxygen consists of two oxygen atoms holding hands, sharing a pair of electrons between them. They don't hold hands very tightly. The O=O bond can be split by ultraviolet light, which constitutes about 10 percent of the sun's radiation. Biological reactions will also split oxygen into oxygen atoms, essential for corralling electrons at the end of the metabolic chains through which aerobes like us extract energy from food. ⸺

Breaking an oxygen molecule's relationship with another oxygen molecule initiates a wild cascade of reactions as the orphaned oxygen atom, now missing an electron that its former friend took with it, searches for a new partner with an electron to share. The orphan—an oxygen radical—is sufficiently electron-hungry that it will accept all manner of company to obtain one. Many of those relationships aren't very stable, so the oxygen radical moves around, displacing other atoms to temporarily pair up, leaving a wake of themselves-now-radicalized molecules that need to keep reacting to return to a stable situation. Sooner or later, oxygen will find its way to two hydrogen atoms and settle down as water, but it may do a lot of damage first.

Aerobic creatures—humans, horses, yeast, grapevines, and so on— manage the daily reality of dangerous oxidation reactions by pairing strategies for accessing oxygen with protection against its rampaging tendencies. That is, we employ antioxidants. Wine similarly needs antioxidants to withstand oxygen exposure without tasting oxidized (see the OXIDATION box), especially since some amount of oxygen exposure is necessary for yeast survival 🍷 and sometimes useful for texture. 🛢 The chain reactions that oxygen trips are tangled, and especially so in wine, where one oxidized phenolic molecule can lead to another and another—a series of events at the core of how wine's astringency and color evolve over time.

Vitamin C is an antioxidant. Phenolics are antioxidants, including the ones that give wine color and character. Sulfur dioxide is an antioxidant. This is an instance in which distinguishing SO_2 from sulfites is important, because only the former will absorb oxygen radicals. Sulfur dioxide is also sulfur's antimicrobial form.

OXIDATION

Oxidation is easy to recognize if you enjoy sherry or one of the handful of other oddball, deliberately oxidized wine styles such as the Vin Jaune produced in the Jura region of France. Their characteristic nutty, bruised-apple qualities come courtesy of acetaldehyde, formed when ethanol is oxidized and abundant in these wines, though considered a flaw elsewhere. In the case of some sherry styles, and Vin Jaune, ordinary chemical oxidation gets a biological bump. *Flor* yeast, the odd breed of *Saccharomyces cerevisiae* that films the surface of these wines when they're deliberately aged in half-full barrels, have extra copies of the alcohol dehydrogenase enzyme that metabolizes ethanol into acetaldehyde. Consequently, it performs this reaction faster than other yeasts. (Acetaldehyde itself is a toxic mutagen that causes breaks in DNA, and that's central to alcohol's toxic effects in humans when it builds up after copious drinking, but *flor* yeasts are unusually resistant to its effects.) *Flor* strains have also accumulated cell-membrane changes that make cells buoyant and hydrophobic. Because they mix poorly with water, *flor* cells tend to clump together in the least-liquid environment they can find, forming a layer at the wine's surface. Beyond this list of unusual characteristics, yeast biologists are still working out the biological background of how *flor* came to be *flor*.

Unfortunately—unfortunately for wine chemistry students, at least—sulfur exists in several other forms in wine too.

SO_2 dissolved in wine toggles among sulfur dioxide (SO_2), bisulfite ions (HSO^{3-}, formed when SO_2 and H_2O react and release an H^+ ion), and sulfite ions (SO_3^{2-}, formed when bisulfite ions lose H^+). Those forms are in equilibrium; the reactions that link them go both ways, so their relative abundance is a function of how statistically likely one molecule is to react to form another. Statistically, a greater abundance of H^+ increases the likelihood of bisulfite and sulfite reacting with H^+ ions to form SO_2. Wine is acidic, which is another way of saying that it's full of H^+. In wine, therefore, most sulfur dioxide does indeed exist as SO_2. However, SO_2 itself toggles between "free" SO_2 and "bound" SO_2, the latter attached more or less temporarily to acetaldehyde or other molecules. For the most part, only the free fraction does the things that people want SO_2 to do.

Free SO_2 is small and uncharged, so it can diffuse through microbial cell walls. ⁑. The inside of a cell is much less acidic than wine, with many fewer H^+ ions. Much of the SO_2 that infiltrates a cell therefore becomes bisulfite. (A small fraction becomes sulfite.) Bisulfite interferes with proteins by interrupting bonds between sulfur-containing amino acids, disabling protein-based enzymes. It also destroys thiamine, vitamin B_1 (which, incidentally, is a nutritional reason to limit SO_2 in foods). And bisulfite can induce mutations in DNA, explaining why sulfite can cause cancer in rodents fed too much of it.

Assuming that you're not a lab rat, the amount of SO_2 that enters a bacterial cell is far greater than the amount your own cells see, which is why you don't experience the same effects from SO_2 in wine or other foods. Our friendly fermentation yeast *Saccharomyces cerevisiae* is also largely immune, thanks to membrane transporters that specifically shunt bisulfite ions out of the cell. Among their other helpful adaptations, wine yeast have evolved to make more of these membrane transporters than their cousins that aren't employed in winemaking. ◗

Only free SO_2 participates in these antioxidant and antimicrobial activities. Consequently, calculating an SO_2 addition involves accounting for how much will bind to other molecules, as well as the wine's pH. This isn't remotely simple. Acetaldehyde is the major binder to consider, responsible for roughly 80 percent of the total. More SO_2 in the environment leads yeast to produce more acetaldehyde. Cooler fermentation temperatures lead yeast to make less acetaldehyde, and also reduces the efficacy of SO_2 as an antimicrobial compared with higher temperatures. But SO_2 leaves wines faster at higher temperatures. SO_2's antimicrobial properties mean that it alters which microbes are metabolically active, which changes how much acetaldehyde microbes produce, which adds even more turns to this knot of interconnected relationships, even before we account for binding molecules other than acetaldehyde.

The very reactivity that makes SO_2 so useful as an antioxidant means that it has other chemical effects too. SO_2 binds to and inactivates browning enzymes. 🍷 It also binds to pigmented anthocyanins, though the resulting color loss or "pigment bleaching" is thankfully reversible and temporary. It temporarily binds to aldehydes and ketones, reversibly dampening their aromas. The ratio of free to bound SO_2 matters in every case. And even when these effects are temporary, they interfere in the dynamics of how these molecules participate in reactions in a wine's early days, ramifying as it ages.

Reduction

When sulfites became a public issue in those salad bar days, some consumer advocates marshaled for a total ban on SO_2 as an added preservative. Had their campaign been successful, most wines would still bear "contains sulfites" warnings. Yeast produces sulfites as a routine part of the business of fermentation—though, as usual for yeasty tendencies, strains can be selected that make more or less such that the occasional wine ends up below the threshold at which labels are required.

Sulfur atoms contribute to several amino acids from which cells build proteins. Yeast (and humans, for that matter) therefore need to ingest it. Yeast releases any unneeded sulfur in the form of sulfide ions, S^{2-}, which pick up two H^+ ions before leaving wine as hydrogen sulfide gas.[10] Hydrogen sulfide is better known as the aroma of rotten eggs (see Figure 10.1). Fortunately, that aroma encounters noses as it's *leaving* the wine, generally well before bottling. Unfortunately, the complexities of sulfur biochemistry mean that it may also dissolve and stick around. When, having initially dissolved, volatile sulfur compounds then emerge from a finished wine, we say that it suffers from reduction (a condition associated, at times, with gas-impervious screw caps). ▌ This is confusing, because when chemists talk about reduction, they're describing a molecule gaining electrons—that is, the opposite of oxidation, or losing electrons. (Oxidized compounds sometimes lose electrons to oxygen, but other molecules can also be

---

[10] For reasons having to do with the intricacies of how yeast respond to stress, they're most likely to offload surplus sulfur when they're also short on nitrogen. 🍷

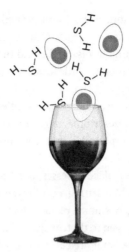

Figure 10.1 Hydrogen sulfide smells like rotten eggs and is one component of the wine fault known as reduction.

involved.) Reduction is typically categorized as a wine fault, but volatile sulfur compounds are also essential to desirable wine aromas.

As is often the case for so-called faults, volatile sulfur compounds are a boon in moderation and a bane in excess. The line between "moderation" and "excess" varies with context, personal sensitivity, and preference. Individual variation is why I have to talk about "so-called faults," since the difference between a winemaking error and enjoyable flavors always needs to be judged in context—the wine's, and the drinker's. By the numbers, the same quantity of a molecule such as hydrogen sulfide may be offensive or imperceptible depending on what other flavors surround and interact with it. And while detractors will claim that natural wine fans are only pretending to like wines with funky, feral flavors that would be patently unacceptable in mainstream specimens, I'll speak for myself in saying that those flavors can evoke deliciously visceral reactions that simple fruit flavors don't. I *still* don't like rotten eggs in my wine.

Above their perception thresholds (which tend to be low; they're pungent), sulfurous compounds such as hydrogen sulfide, methanethiol, and dimethyl sulfide can be unpleasantly reminiscent of rotten eggs, rubber, onion, sewage, and canned vegetables. They may also dampen fruit and floral aromas. Others, including a whole class of fruity sulfur-containing thiols, are behind the passionfruit notes of Marlborough sauvignon blanc and the struck-match quality of some Chardonnay-based wines. 🐚 In below-threshold quantities, all of the above will contribute to overall flavor in subtle and synergistic ways. "Reduction," as a wine descriptor, refers to having too much of the wrong kind of sulfur compound in the wrong place, not the mere presence of sulfur compounds.

Assuming that out-of-place excesses aren't involved, retaining varietal characteristics while letting go of swamp gas requires giving highly volatile hydrogen sulfide plenty of room to escape without introducing enough oxygen to destroy less volatile, rather delicate, desirable compounds. The latter are easily oxidized, which is one reason why aromatic white wines tend to be given more sulfur than reds, to protect their aromas. (Reds also have more built-in antioxidant protection thanks to their extra skin- and seed-derived phenolics). Adding sulfur dioxide early in winemaking translates to more volatile sulfur compounds later, between yeast-sulfur metabolism and how sulfur compounds react, all of which contributes to how sulfur shapes wine flavor beyond simply preserving it.[11] You're probably getting the idea by now, if you didn't have it before, that very little about SO$_2$ could be called simple.

Using Less

Winemakers on the whole are using less sulfur than their predecessors—not only because of wary drinkers or labeling laws (see the ORGANIC box), but because biochemists, microbiologists, and the industry at large keep learning more about how SO$_2$ works. For example, because SO$_2$ provokes yeast to produce acetaldehyde, adding it at the start of a fermentation paradoxically makes it necessary to use more, because more of the total will become bound up

---

[11]   Lyu et al., "Effects of Antioxidant and Elemental Sulfur Additions."

ORGANIC

Wines labeled as "organic" in the United States not only must be made from organic grapes, but also must be made with no added sulfur. Wines bearing the less intense moniker "made with organic grapes" can contain sulfur, but must be made with grapes from vineyards with a current organic certification, an expensive and intensive process that requires adherence to one among multiple interpretations of "organic." Consequently, many excellent wines are made with grapes that might otherwise be thought of as organic—carefully and sustainably grown—without saying "organic" on the label.

and thus unavailable to do its job. Similarly, while grapes affected by rot might seem to require more sulfur to tamp down the load of spoilage microbes they carry, damaged grapes also come with more SO_2-binding compounds. The result can become an arms race, with winemakers making bigger sulfur additions to achieve a target concentration of free SO_2, which increases the proportion of bound SO_2, which increases the sulfur they need to add. Or, appreciating this bit of biochemistry, they can short-circuit this spiral by removing sulfur-binding compounds with extra pre-fermentation clarification.[12]

Attitudes toward preservatives have also shifted, so sulfur is now more likely to be a strategic tool rather than unthinking insurance against spoilage. Palates have changed too. American and Australian drinkers today like their wines far less sweet, on the whole, than they did a century ago. Residual sugar in a finished wine is an invitation to microbes to come spoil it. Dry wines require less antimicrobial protection and are less likely to go bad.

Winemakers also now have substantial physical control over how much oxygen their wines see, from blanketing juice with inert gas to exclude ambient air, to precision bubblers that infuse specific doses of

---

[12] Jackowetz, Erhu, and de Orduña, "Sulfur Dioxide Content."

oxygen into a tank, 🛢 to "technical" closures that let air into a bottle at a defined rate. ❙ Even improved transportation is a factor. When grapes were shipped in rail cars that might take days to reach their destination, spoilage could happen before they even arrived at the winery if grapes weren't covered in sulfur.[13] 🫒 Now, refrigeration works wonders.

Recent studies recommend that microbial growth during fermentation be "stabilized" with less than half the sulfur that winemakers were instructed to use in the mid-twentieth century. What's more, sulfur dioxide itself doesn't seem to be the key factor in averting spoilage. The most important consideration in whether spoilage microbes thrive, research suggests, seems to be the degree to which desirable microbes have the opportunity to exclude them. Adding commercial yeast is one way to do that. Adding SO_2 is another. A third is cultivating a robust community of diverse microbes who tend to resist intruders. 🍇 Leaving finished wines with a dearth of nutrients so that unwanted bugs have little to eat goes a long way too.

(The Impossibility of) Finding Alternatives

Those developments lead to a place where lower doses of sulfur can be used less automatically and more judiciously. At the same time, wine technologists have been searching for alternatives that might offer a sense of antioxidant or antimicrobial security similar to what a robust dose of SO_2 provides. While they're a varied bunch, they all come with the same problems: they don't mimic sulfur's spectrum of actions, have their own downsides, and may raise just as many concerns for additive-wary consumers.

Chitosan is a leading option because it's both a strong antioxidant and a strong antimicrobial. It's also the second-most-common polysaccharide in the natural world, right behind the cellulose that makes wood woody, because it's the main structural component of fungal cell walls and of insect and crustacean exoskeletons. Only chitosan from fungi is authorized for winemaking use, so folks with shellfish allergies needn't be concerned. However, if a consumer is concerned about

---

[13] Winkler and Jacob, "Utilization of Sulfur."

"additives" generally, chitosan is no help at all. And, having different properties, it's not a one-to-one replacement for sulfur dioxide.

The same can be said for dimethyl dicarbonate or DMDC, which acts against microbes by damaging enzymes essential for energy production. DMDC's overwhelming advantage is that it's quite effective against *Brettanomyces*, which can sometimes tolerate SO_2. 🐌 DMDC's disadvantages are just about everything else. It will kill *S. cerevisiae*. It's expensive. (SO_2 is cheap.) It's flammable, toxic, and poorly soluble in water, so applying it requires special machinery. Though it rapidly breaks down in wine, it decomposes into methanol and carbon dioxide. Even if a small methanol addition would rarely raise a wine above safe and legal limits, 🐌 adding any quantity of a known toxin is hardly a selling point. At the end of the day, DMDC is mostly used in sweet wines (its only approved use in the EU) for which SO_2 may not suffice.

Some alternative antimicrobial strategies disable microbes by mechanical means. Wine can be pressure-treated, microwaved, zapped with high-intensity electric fields, irradiated with UV light, subjected to intense ultrasonic vibration, or pasteurized. These have the advantage of not adding anything directly, but they leave traces by changing wine components themselves. High pressure snaps weak associations among molecules and accelerates reactions, effectively pushing pressure-treated wine to age faster. Microwaves and ultrasound agitate cell membranes and kill microbes through tiny pockets of intense heat. They're also so good at producing aged characteristics that they're occasionally used expressly for this purpose. In the worst-case scenarios, treated wines come away tasting cooked.

In a different vein entirely, wine is seeing a shift from chemistry to biology, a move to replace chemicals with specially selected microbes. Generally-recognized-as-innocuous bacteria and yeast may be inoculated to eat up nutrients that could otherwise feed spoilage microbes. Bacteriophages—viruses that infect specific classes of bacteria—might suppress unwanted bacteria while posing zero threat to yeast or humans.[14] Viruses don't produce their own flavors, but bacteria- or yeast-based

---

[14] Bacteriophages have no negative effects on humans whatsoever. However, they run the serious risk of spreading to places where the bacteria they target are desirable. Phages that kill lactic acid bacteria, for example, are a real danger for cheesemakers.

"bioprotection" can come with fringe benefits, wanted or not, in the form of a wider range of flavorful microbial metabolites. What they don't come with are SO_2's other activities. Biology for chemistry is, once again, never a one-to-one replacement proposition.

The same shift from chemistry to biology is happening widely across agriculture and industry. Farmers who have relied on heavy-duty nitrogen fertilizers since the 1970s Green Revolution are coming to see that badly depleted soils have become little more than a physical matrix for supplements—and that this is terrible for long-term agricultural sustainability. A growing number are feeding fields with live microbes in hopes of rebuilding soil ecosystems through which crops will make better use of nutrients in the first place. In a different vein, industries that have relied on petrochemical inputs—plastics, textiles, industrial chemicals that end up in all kinds of products, and so on—are shifting, with more or less verve, toward *fermenting*. Synthetic biologists, who use molecular biology techniques to redesign DNA, are teaching yeast and bacteria to produce an endless array of valuable molecules from whatever sugar source might be available to fuel them.

Thinking of "biologicals" as "greener" than chemistry is almost inevitable, and by design, but nothing comes without its costs. In the case of engineering microbes to ferment chemicals, the sugar that feeds them still has to come from somewhere. "Somewhere" may be a monocrop such as corn grown in place of food or forests, which isn't great. A challenge now is to slot waste products from other industries (maybe including wineries) into these processes to reduce the number of new problems caused by trying to resolve old ones, bringing industries closer to circular recycling and further from an endless game of whack-a-mole. 🐰

I'm not sensitive to sulfur and have nothing against it on principle. I also enjoy flavors that come from using very little of it. Otherwise identical wines made with different doses of added sulfur have distinct molecular fingerprints even after years of bottle-aging.[15] Alternative strategies leave their own unique signatures. Low- and no-added-sulfur wines that rely on the internal balance of microbial communities

---

[15] See, e.g., Roullier-Gall et al., "Sulfites and the Wine Metabolome."

are often a world apart from grape-forward, freshness-focused, thoroughly sulfited counterparts. To my palate, fresh and fruity becomes boring after a while. But most of all, it's wonderful to see wineries experimenting with using less SO_2 in the interest of seeing it as a tool, a set of stylistic choices to be controlled, rather than simply used "safely." The result is far more interesting than a debate with two sides.

11

Sugar 🍇

My local wine store, and probably yours too, shelves their wines into sweet versus everything else. One might reasonably imagine that everything that isn't a sweet wine is a dry wine. One might also reasonably imagine that the sweet wines contain notable quantities of sugar and the dry wines don't. As we'll see, neither of those statements is true or helpful. Ostensibly "dry" wines often aren't. And sweetness isn't just about sugar. But first, we need to start with yeast.

Our Friends the Sugar-Eaters

Saccharomyces cerevisiae means "sugar-eating fungus," though its preferences are a bit more discerning than that moniker suggests. If you're a member of this species, your favorite food is glucose, the preferred fare of many fungal and most animal cells. The cell membrane that separates you from the rest of the world is riddled with transporters to import this sometimes-scarce resource when you find it. 🍇 If glucose is super-abundant, however, you have a different problem: water. You need water to survive even more than you need glucose. Glucose and other sugars, in their unthinking chemical way, are greedy with water. Water and sugars both bear relatively strong charges. Opposite charges attract, such that a sugar molecule will hold onto a halo of water molecules by way of electrostatic attraction if given the opportunity. This hydrophilicity is why table sugar dissolves easily in your tea. It's also why concentrated sugars are toxic to our little yeast cell, which needs the water that sugar refuses to give up. If you've ever made jam, you've deliberately employed this property to keep microbes from eating your food. Preserves preserve fruit because even though they're moist, they contain so much sugar that there isn't enough water

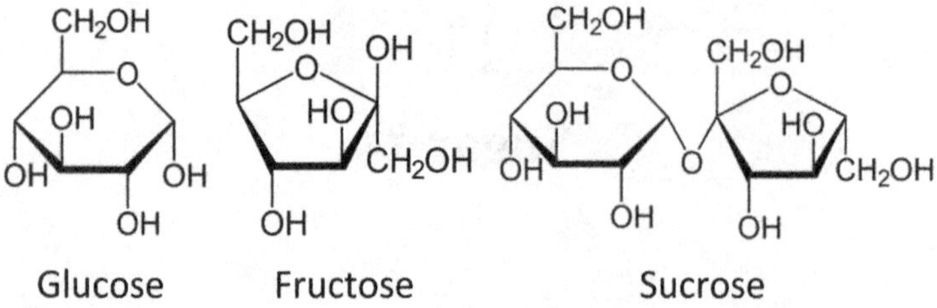

Glucose　　　Fructose　　　　Sucrose

Figure 11.1 The three sensory-relevant sugars that might be residual in wine.

available to microbes in the environment presented by a jam jar for most bacteria and yeast to grow.[1]

Saccharomyces's dietary preferences are predictable: glucose first, then fructose (see Figure 11.1). Grapes contain approximately equal quantities of these two sugars. That's a bit odd, because plants' preferred energetic currency is a third sugar, sucrose—a disaccharide made from one molecule of glucose bound to one of fructose. (Sucrose is also the chemical term for the table sugar that you may have used to sweeten that jam.) However, grapes contain enzymes that split the plant's sucrose into its component parts during ripening. Those enzymes will similarly split any sucrose that a winemaker might add before fermentation to bolster a wine's alcohol content, making its component parts available for yeast to digest.

Grapes also contain some pentoses, which have five carbons in contrast to glucose's and fructose's six. Yeast won't touch those at all. As it turns out, that doesn't matter much to perceived sweetness because pentoses don't taste very sweet. The residual sugars we *do* care about in wine are glucose and fructose, and two factors are behind whether a finished wine contains them: either the winemaker actively prevents yeast from finishing its job, or the yeast quit early without the winemaker's permission. Before we discuss either of those scenarios, we should discuss residual sugar itself.

---

[1] Modern fruit preserves, built for an era of ubiquitous refrigeration, often contain less sugar than traditional recipes and so are more prone to spoil if you leave them out on the tea table.

Residual Sugar

Residual sugar is usually abbreviated RS, but that acronym was originally used for *reducing* sugars, after the method used to measure them. In chemistry argot, a molecule is *reduced* when it gains an electron from another molecule, which is called *oxidized* when it loses that electron. Reducing sugars are inclined to reduce other molecules. Because glucose and fructose are reducing sugars, a convenient way to find them is to give them something to reduce—ideally something such as a copper ion that changes color (from bluish to yellowish in this case) when it receives that electron. That convenient bit of chemistry means that an easy test will yield a rough evaluation of the combined glucose and fructose left in a wine. Inconveniently, though, other wine components such as tannins can also reduce copper, making the results of that test a bit squishy.

According to the Organisation Internationale de la Vigne et du Vin (OIV), which legislates for the vast majority of the winemaking world, that imprecision means that everyone should stop measuring reduction and should instead measure glucose, fructose, and sucrose directly.[2] (Leaving out pentoses makes little practical difference because they're substantially less sweet than sucrose, and they're not abundant.) Because measuring sugars requires more expensive equipment, wineries with less cash for lab updates may still use the old method. Those two methods can disagree to the tune of up to several grams of sugar per liter, such that residual sugar reported on a tasting sheet or a back label won't reliably mean the same thing from winery to winery. That's obviously confusing, but as we'll see, it proves to be pretty low on the list of complications involved in separating wines into "dry" and "sweet."

No matter how you document them, residual sugars in finished wines are typically weighted toward fructose—which humans perceive to be about twice as sweet as glucose—because yeast eat the glucose first. Sucrose only appears if a winemaker has added it after fermentation. That's uncommon outside of fortified and dessert wines. Table wines boasting "no added sugar" are a bit like "gluten-free" stickers on

---

[2] Wine-producing countries choose whether or not to subscribe to and be governed by the OIV. Most have. The United States, Canada, and China haven't.

bags of chopped broccoli—they're true, but the likelihood of that dec-laration *not* being true for that kind of product is slim.

That said, direct sucrose additions aren't the only way to bolster a wine's sweetness. Sugar can be added in the form of unfermented grape juice, just as "no sugar added" granola bars rely on fruit-derived sweeteners.[3] Alternatively, fermentations can be stopped on purpose be-fore they're finished, whether by killing yeast with a large dose of sup-plemental alcohol or meticulously filtering out all of the microbial cells. Either is usually combined with a severe chilling that effectively puts yeast to sleep, plus some powerful preservatives. Pasteurizing is another (al-beit less common) option. One might even intentionally give yeast more sugar than they can handle before they succumb to alcohol poisoning. Intentional yeast alcohol poisoning is the mechanism behind ice wines, botryrized wines like Sauternes and the sweetest styles of riesling, and wines made from partially dried grapes such as Recioto della Valpollicelli. All involve concentrating grape sugars by removing water via freezing, fungus-induced shriveling, or straightforward dehydration, so that plenty of extra-sweet fructose remains when yeast reach their limit.

Alcohol poisoning sometimes also happens in the big, fruity alcohol bombs that California in particular has become known for making from ultra-ripe grapes. Commercial yeasts have been cultivated to with-stand more and more ethanol—as much as 17 or 18 percent, more than you'd want for stylistic and drinkability reasons even outside of other considerations. If, however, you're fermenting with ambient yeasts that reside in your winery on their own terms, 🫘 you might have the poor fortune to work with a strain that can only tolerate 15 percent alcohol in a zinfandel that would otherwise have reached 16, leaving residual sugar out of keeping with the style you might have wanted.

Excepting exceptional circumstances when yeast are stopped on purpose, as in ice wine, the best-case scenario for both yeast and wine-maker is that *S. cerevisiae* continues fermenting until essentially all the glucose and fructose are gone. Unfortunately, that isn't always what happens. Sometimes, yeast get stuck.

---

[3] Dates, apple juice concentrate, honey, and the rest may contain a different break-down of specific sugar molecules, but in the end, sugar is sugar, and eating a date means consuming lots of it.

Sticky Situations

"Stuck fermentation" is the general term to describe a fermentation that proceeds until it doesn't. At the heart of many such sticky situations is that yeast need more than sugar and water to thrive. Growing and multiplying require other essential nutrients that must be taken up from the environment. Grape juice is replete with micronutrients, so this often isn't an issue, but nitrogen can be in short supply. Despite being a necessary component of proteins and amino acids and other biological molecules, much of the nitrogen kicking around in grapes isn't in a form that yeast can use. Winemakers therefore routinely add extra yeast-assimilable nitrogen, YAN, so that nitrogen isn't a limiting factor. Scads of research have been aimed at characterizing just how much YAN a yeast cell needs, because adding more than that leaves surplus nutrients sitting around to feed spoilage microbes. No one wants that.

Other nutrient insufficiencies and imbalances work the same way, even if they're not always so obvious. Yeast ingest the potassium they need by exchanging a potassium ion outside the cell for a hydrogen ion from inside. That simple mechanic leads to complex relationships between a yeast's potassium needs and a comfortable pH.[4] Those two factors can occasionally be so far out of whack with each other that fermentation stops.[5] Oxygen also counts as a nutrient in this case. Even though it isn't necessary for metabolizing sugar into ethanol, yeast need oxygen to manufacture the membrane components that let cells survive alcoholic environments. ❧ In the presence of a lot of sugar and the absence of enough oxygen—as might happen, say, in the depths of an inadequately mixed fermentation tank—yeast will get stuck.

The potential limitations of nitrogen, oxygen, potassium, and numerous micronutrients are well established. The recent surprise in sticking research is that inter-microbial relations may also be at play. Profound bacterial growth in serious spoilage situations can

---

[4] pH is casually used as a measure of acidity—lemon juice has a low pH, lye a high one—but it's formally a measure of the quantity of hydrogen ions in a solution. Acids are defined by their propensity to release hydrogen ions, and bases by their propensity to absorb them, so an acidic solution is an H^+-rich environment.

[5] Kudo, Vagnoli, and Bisson, "Imbalance of pH and Potassium."

outcompete yeast for resources, and some bacteria produce yeast-inhibiting byproducts, including acetic acid bacteria that metabolize alcohol into vinegar. Many strangely stuck ferments don't have this easily detectable problem. Instead, at least some appear to be suffering from something like a case of mad cow disease.

What stuck ferments and mad cow disease share are prions, small proteins that trigger other proteins to fold in dysfunctional ways. Prions have made the news when humans inadvertently ingest them by way of some bad burger, but some bacteria appear to make a different (utterly harmless to humans) set to feed to yeast on purpose.[6] These bacteria don't want alcoholic environs and do want some of the glucose that yeast otherwise hog. To engineer an environment better suited to their needs, they secrete a prion that switches off yeast's inclination to consume glucose and only glucose even when multiple food sources are available (see the PRIONS box).[7] Yeast diversify their diet,

---

PRIONS

Finding prions behind at least some stuck fermentations is interesting on a practical level because wine scientists might be able to do something about it. It's also interesting to microbiologists as evidence that some microbes *manage* other microbes for their own benefit. The first hundred years or so of modern microbiology were primarily about studying individual species in isolation. Imagine characterizing the behavior of gray squirrels, or ospreys, or teenage boys by putting a large group of them in a room by themselves, observing what they did, and assuming this to represent typical behavior for the species. That's effectively what microbiology has historically done. Studying microbes in mixed groups is leading to realizations that they're far more social than has previously been appreciated.

---

[6] As a responsible science communicator, I need to point out that beef production is now very, very carefully monitored, so the chances of ingesting disease-inducing prions with your protein is pretty nearly as close to zero as non-zero chances come by.

[7] Jarosz et al., "An Evolutionarily Conserved Prion-Like Element."

which isn't necessarily to their detriment. The bacteria end up with more glucose and less alcohol. Everyone lives happily ever after, except the winemaker, who's stuck with a stuck fermentation and almost surely not happy about it.

Sugar ≠ Sweet

Sugars in grape juice and wine are often measured in a unit formally known as degrees Brix, or °Bx for short. One degree Brix equals one gram of sucrose (or the equivalent in glucose or fructose) per 100 grams of the total solution. Grape juice can come out of the press at 25°Bx, with roughly 25 g of sugar in every 100 g of juice, and sometimes quite a bit more.[8] For comparison, orange juice sits somewhere in the neighborhood of 11°Bx. Officially, no wine has absolutely zero sugar, though many have so little as to register as 0°Bx—less than 0.2 percent, 2 grams of sugar per liter or not quite a third of a teaspoon per 750 mL bottle, which is more or less the limit of what a human can detect. What any given taster can detect in any given wine, however, varies widely with the wine's chemical context.

Acidity makes sugar taste less sweet. That's hardly news to those of us who eat and drink things, but the interesting point is how that element of perceptual blending is built into wine styles. Riesling, a grape that can be painfully rich in acid, is the paradigm example. If you enjoy riesling—you should; it's wonderful—you may have seen the International Riesling Foundation's helpful sweetness scale decorating a back label. The scale is designed to help consumers choose bottles without needing to master the arcane German categories that earned the variety a regrettably tenacious reputation for being difficult to understand, too sweet, or both. And yet the scale reports not on sugar but on sugar-to-acid ratios. A wine with 8 grams of sugar and 9 grams of

---

[8] No one is out there weighing the sugar in their juice. In practice, degrees Brix estimated via specific gravity—the relative density of a solution compared to water—measured either with a refractometer that visualizes how much light bends when it moves through a solution or, alternatively, with what amounts to a ruler attached to a buoy that can be floated in a vat or tank.

acid is classed as dry. So is a wine with 3 grams of sugar and 4 grams of acid. Both are entirely plausible dinner pairings before dessert makes an appearance, even though the latter contains a third as much sugar as the orange juice you may have drunk with breakfast.

In another example of perceptual blending news that will surprise no one, sugar also makes bitter compounds taste less bitter and makes astringent compounds feel less astringent, ⮥ which explains why folks spoon sucrose into their coffee and their tea, respectively. If you make cocoa from scratch at home, you may find yourself adding less sugar to the cup of a family member who lives on kale and endive, and more for the person who would rather go hungry than eat a salad of bitter greens. An important distinction from a sensory science perspective is that sugar seems to *cancel out* bitterness, making bitterness more tolerable because we perceive less bitterness, rather than distracting us from the bitterness that we're still perceiving just as strongly. Incidentally, another reason why Riesling so often benefits from a bit more sugar than other varieties is that it can be high in phenolics, which sometimes bring bitterness to white wines. 🛢

The connection between sugar, tannins, and the roughness tannins produce is less well described, but one theory centers on the proteins that typically keep your saliva slick and your mouth lubricated. Tannin molecules bind to proteins in clumps that aggregate, becoming larger and larger until they fall out of solution. Astringency seems to be both a chemical sensation and a physical one, the sensation of having a mouth full of little proteinaceous pebbles combined with a corresponding drop in the viscosity of the saliva that ordinarily keeps everything in the oral cavity from sticking together.[9] Tannins bind to proteins with less intensity in the presence of sugars, probably because sugar molecules physically get in the way.[10]

Beyond acid, bitterness, and astringency, tiny quantities of sugar that very few people are ever likely to perceive as sweet have other, more holistic effects on flavor. Research for Alaska Airlines' in-flight beverage program, for example, found that a mere 0.4 g more residual sugar made a white wine less angular and more likeable at cruising

---

[9]   Schöbel et al., "Astringency Is a Trigeminal Sensation."
[10] Rinaldi, Gambuti, and Moio, "Precipitation of Salivary Proteins."

altitudes.[11] That's one more reason why *perceived* sweetness needs to be distinguished from the chemical business of quantifying sugars, because perceptions are affected by what else in addition to sweetness is competing for our sensory attention.

Sugar also isn't the only thing that tastes sweet. Glycerol, a major byproduct of fermentation, is only about half as sweet as table sugar, but contributes both to sweetness and to viscosity, the weight of how wine sits in the mouth.[12] Alcohol does too, which becomes significant to the mouth-filling juiciness of those fat 15 percent cabernets. So does diethylene glycol, which became significant to Austria in the 1980s. The Austrian wine industry still hasn't quite lived down the scandal that came to light in 1985 of some winemakers adding the active, toxic ingredient in antifreeze to underripe wines in pursuit of sweetness and body. (The country's wines are now excellent and unquestionably unadulterated, thanks to exceptionally tight quality control in the interest of never having what happened in 1985 happen again.)

Dissolved carbon dioxide—less than the amount anyone would detect as carbonation—increases a wine's perceived sweetness, even though it doesn't taste sweet on its own, and even though CO_2 is weakly acidic when dissolved in water.[13] Specific protein components derived from dead yeast taste sweet, possibly explaining why some wines seem sweeter after sitting on their lees for a few months. Quercotriterpenoside, a mouthful of a molecule found in oak, similarly explains the observation that wines tend to sweeten up with oak aging.[14]

Another reason why oak may make wine taste sweeter is that oak often contributes vanilla flavors. Humans are so very good at integrating discrete molecular sensory inputs into a unified taste sensation that we easily confuse sweetness with flavors that often travel with sweetness but aren't themselves sweet. Fruity wines often seem sweet

---

[11] Perdue, "Making Wine."
[12] Glycerol is also known as glycerin, the name it usually bears as an ingredient in bubble-blowing solutions.
[13] Gawel et al., "Effect of Dissolved Carbon Dioxide."
[14] Marchal et al., "Influence on Yeast Macromolecules on Sweetness"; Marchal et al., "Identification of New Natural Sweet Compound."

even when they contain no residual sugar, because most fruits that we think of as quintessentially fruity are sweet.

All of this boils down to the difficulty of categorizing wine into "sweet" and "dry." I used to think that "dry" was a terrible wine word. It can't be cleanly contrasted with "sweet" for all of the reasons we've touched upon. Worse, some casual drinkers will use "dry" as a proxy for "good," sometimes because they think they *shouldn't* like sweet wines, while others will use "dry" as a proxy for "bad," because their main exposure to the vinous world has been overtly semisweet table wines.

I've since changed my mind. What I now think is that, like minerality, "dry" is a useful descriptor not because it correlates to a specific molecular reality, nor because it unambiguously means the same thing to everyone in every context, but because it conveys an impression about an experience. Even though the International Riesling Federation's sweetness rating is principally calculated via sugar and acid, it is always confirmed with a taste test. "Dry" says something about the experience of tasting without suggesting that pinning that experience on a specific molecular mechanism is necessary to justify it.

A detailed readout of a wine's chemical composition is not a useful tasting note. That's true because wine chemists have yet to map all of the intricate details of how molecules synergize to yield sensory perceptions, *but also* because different people perceive different sensations when confronted with the same molecular mixture. Flavor, in any kind of meaningful way, is always the product of a taster tasting, which means accounting for the humans in the equation. The tasting panel that confirms International Riesling Federation sweetness scale is calibrated to know exactly what they're looking for. Winemakers and other professional tasters tend to be more attuned to detecting residual sugars than the average drinker, ♪ so a winemaker's sweet is routinely a consumer's dry. Anecdotally, habitual German wine drinkers, who might drink more riesling than the global average, accept a bit more sugar in their table wines than equally sophisticated consumers in the United States. In short, "dry table wine" on the label of a red blend at your local shop is an indication, not a guarantee. For all the consternation that can cause at dinnertime, it's not a problem to be solved simply by looking for a percentage of residual sugar on a back label—no matter how a winery is measuring it.

12

Oak

Pie is a storage solution. When communal ovens and eating lunch in the field were in vogue, a sturdy flour-and-fat paste could encase your stew for baking, make it portable, and (when the crust was edible) funnel some extra calories into the eater too. Thinking of crust as containment led to the possibility of pies for entertainment instead of eating. The four-and-twenty blackbirds from the nursery rhyme starred in just one example of a whole genre of special-effects pies designed for the pleasure of the nobility, with any number of amusing fillings: songbirds, other live animals, even human entertainers. (Crusts were assembled after baking with fillings that wouldn't survive a trip through a hot oven.) As hours-long banquets punctuated with practical effects fell out of fashion, cooks switched their focus to palatability alone, and today we encounter little else.

Oak barrels are like pie. Oak is also an anachronistic container for which we've developed a taste. I doubt that many present-day members of traditional pie-eating cultures need the extra calories pie crust provides, nor are we likely to make pie for convenience's sake; the opposite is more true. And yet I own a pie plate, and maybe you do too, because pies are well loved. So, for that matter, are oak-forward wines. Buttery, vanilla-flavored chardonnays, the product of extensive oak contact (malolactic fermentation also contributes to buttery notes 🍾), are among the best-selling wines in the United States. They're also widely despised by critics for being boring, sweet, one-note butter bombs. The word "vulgar" sometimes makes an appearance, whatever that means these days.

Barrels may not be expensive calorically, but they're expensive in just about every other way. Their round shape, a function of how coopers make wood watertight, eased transportation when they were literally rolled from place to place; now, their roundness makes loading a forklift precarious.[1] They demand intensive cleaning and considerable maintenance. They easily become reservoirs for unwanted microbes. They're pretty, but that hardly seems like a fair trade. And yet they remain among the most widespread of winemaking tools, routinely described as working magic on the wines they hold.

Barrels' seemingly magical effects—on texture and color, more than flavor as such—stem from their ability to microdose wine with oxygen, plus oak's richness in powerfully reactive phenolic compounds. Put differently, while the original value of a barrel may have stemmed from being watertight, its long-term value is a function of not being solid at all. A good barrel won't leak wine, but it will leak gas. It's also full of compounds just waiting for an opportunity to dissolve and join a liquid. That's not magic; it's physics with a side of chemistry, and it starts with diffusion.

Diffusion

In the mid-nineteenth century, a German physician named Adolph Fick set out to explain how oxygen moved from air into the blood and ended up describing diffusion. Fick recognized that while air and blood never mix directly in the lung, oxygen (and carbon dioxide) could move through the capillary walls that separated the two. What we now call Fick's Law is an equation for calculating how fast gases move between two compartments. Thicker walls slow diffusion down. A larger difference between the concentration of the gas in question on one side of the wall and on the other—a larger concentration gradient—speeds diffusion up.

Wine in a barrel is like blood in a pulmonary capillary, only thick barrel walls breathe more slowly and wine consumes less oxygen than we do. Ambient atmospheric air is at play in both cases, containing approximately 21 percent oxygen.[2] Depleted venous blood coming in from

---

[1] Empty, the most common 225-liter Bordeaux *barrique* weighs about 100 pounds or 46 kilos.

[2] At sea level, air contains about 20.95 percent oxygen. Altitude doesn't change the *proportion* of O_2. However, the effective *quantity* of oxygen to which aerobically respiring creatures—and wine barrels—are exposed decreases as atmospheric pressure decreases. At my desk in Colorado at 5,000 feet, my lungs see a quantity of oxygen equivalent to 17.3 percent of ambient air were I dipping my toes in the ocean.

a body's extremities and the wine inside a barrel both contain very little oxygen; whatever they contained has already been consumed. An adult human at rest consumes something on the order of 3.5 milliliters of oxygen each minute for every kilogram they weigh.[3] Because the side of a barrel is roughly 2.2 centimeters thick while lung capillary walls are a fraction of a micrometer —and because wine consumes less oxygen than your metabolic equipment does—diffusion through a barrel tops out at around one-tenth of a milliliter of oxygen per liter of wine per day.

In theory, nitrogen and carbon dioxide can move through a barrel just as easily as oxygen can. In practice, they don't, because only oxygen reacts with wine to keep recreating a meaningful concentration gradient. This reactivity is what we mean when we talk about wine *consuming* oxygen, because the standard O_2 gas molecule is split and incorporated into new molecular assemblies. In the first instance, iron or copper ions in wine steal electrons from the bond that holds two oxygen atoms together.[4] The resulting oxygen radicals—the "free radicals" that antioxidant products warn against—then pick up hydrogen atoms to form hydrogen peroxide.[5] Hydrogen peroxide reacts with any number of the varied phenolic compounds that give wine so much of its character, altering color, texture, and flavor in ways associated with increasing maturity.

The oxygen diffusion rates that I cited a moment ago are only ballpark figures. Asking how much oxygen a barrel transmits is like asking how much air a person can hold in their lungs; the answer depends on what kind of individual we're talking about. Larger barrels have lower surface-area-to-volume ratios and deliver less oxygen per liter of wine they hold. Assuming equal volumes, the next most important consideration is the kind of oak from which they're constructed. Wine barrels originate in a limited range of forests across France, the eastern United States, Hungary, and bits of neighboring Romania. As a rule, French oak is more porous than American white oak, because the latter has a relative abundance of tyloses, molecular dams that protect a damaged

---

[3] The exact number is a function of age, fitness level, and other individual variables.

[4] Light will also initiate this set of reactions, though its contribution inside a dark barrel is insignificant.

[5] Hydrogen atoms are always in plentiful supply in wine, since all wines have a relatively low pH and pH is nothing but a measure of hydrogen atoms; lower pH, more H^+.

tree against fluid loss.[6] When the wood is cut, tyloses seal channels through which either liquids or gases might otherwise flow. Then, within any given species, permeability can vary by as much as twofold across the staves that a cooper might select.

The permeability of individual barrels varies too, not just because of differences in staves but because of differences in construction. Extremely tight barrels generate a gentle vacuum as gas molecules are consumed by the wine, so the pressure differential draws in more oxygen than Fick's Law would otherwise predict. New barrels are more permeable than used ones because wine saturates wood over time, occupying pores through which oxygen would otherwise travel. And no matter how meticulously barrels are cleaned between uses—as they must be to hold back spoilage microbes—wine deposits tartaric acid crystals and other sediments that further block those pores.

Exactly *how* oxygen moves into a barrel remains a matter of some debate. Described in general, a barrel is one unit: a barrel. Described in detail, a barrel is a set of staves forced together with galvanized steel hoops, topped and tailed with wooden heads, with a bunghole cut in the side to access the contents. Described in even more detail, a barrel's parts include the gaps between staves and an interface between the bunghole and its stopper. Those detailed pictures make it possible to imagine multiple potential routes for oxygen entry. Dry oak itself holds some oxygen (about 0.3 milligram O_2 per gram of wood) trapped in the wood's pores. That measure will be released into a barrel's first fill of wine fairly quickly. Oxygen also reliably flows through the bunghole whenever the barrel is opened. Beyond that, the relative significance of seepage through minuscule gaps between staves, diffusion directly through the wood itself, or bunghole-related air exposure has yet to be satisfactorily settled.

Micro-Ox

All of this variability aside, if a barrel is tight enough to hold water, it's a tool for micro-oxygenation, or micro-ox. Micro-ox stands in contrast

---

[6] "American oak" in this context means a small number of native white oak species among the many, many kinds of oak that shade North America. That a wider range of white oak species isn't used is probably a historical accident more than anything. American red oaks, on the other hand, rule themselves out as totally unsuitable by being coarse in both grain and flavor.

MACRO-OX

Careful macro-ox ensures that yeast have sufficient oxygen to survive the stresses of a full fermentation. Yeast can get by with no oxygen at all in the short term, but need at least some to sustain multiple generations in alcoholic environments. ⌣⌣ Macro-ox can also be deployed to force molecules that are going to oxidize to do so right up front, notably including polyphenol oxidase or PPO, a grape enzyme that only becomes active in the presence of oxygen. Active PPO oxidizes phenolics. Oxidized phenolics turn brown, which sounds awful, save that brown, oxidized phenolics precipitate and fall to the bottom of the tank, so wines that turn brown in this way don't stay brown for long. Provoking that process early leaves far less potential for uncontrolled, unexpected browning later. Precipitated phenolics can also eliminate unwanted bitterness. The downside is that oxygen also initiates a substantial, sometimes undesirable set of other reactions—breaking up, for example, aromatic thiols responsible for the varietal punch of sauvignon blanc and other aromatic whites. Style considerations related to oxygen's varied activities are why wines may be made "oxidatively" with lots of oxygen exposure, "reductively" with careful anti-oxygen protections (blanketing grapes with CO_2, fermenting in closed tanks, using pumps that don't draw in outside air, etc.), or anywhere along the spectrum in between.

to macro-ox(ygenation). Deliberate macro-ox involves saturating a wine or soon-to-be-wine with as much oxygen as it will hold (see the MACRO-OX box). Any more oxygen exposure is usually an accident, since that way leads toward oxidized characters, characteristic of sherry and a very few other unusual wine styles but considered faults everywhere else. ⬤

Micro-ox as a deliberate, barrel-independent practice is a recent invention, happened upon in the early 1990s as a technology to soften tannins in rough young reds whose planned price point is too low to justify a trip through a barrel. Thanks to digital controls and several decades of experimental data on oxygen management, micro-ox can now be "dialed in" to achieve stylistic modifications in large-scale wine recipes.

Micro-ox can also be applied at any time, not just when wines would typically be barreled, to generate formerly inaccessible effects. All the same, micro-ox often is engaged to mimic barrel-aging without shelling out for an expensive, high-maintenance bit of artisan woodworking.

Replacing a barrel with oxygen-infusing machinery isn't as simple as plugging a few numbers into Fick's Law and solving for X—not only because barrels vary, but because wines vary. Wines range massively in phenolic composition, and therefore in their capacity to consume oxygen, and therefore in how much oxygen they need to consume to achieve a particular stylistic change. Big, tannic reds take far more than delicate, aromatic whites, but the details remain sufficiently intricate and multifactorial that dosing any individual wine can't help but be something of a guessing game.

Soluble Sensory-Active Molecules

Micro-ox can't account for the second major element of a barrel's value: components of the wood itself that dissolve into wine from the barrel's inner surface. Mimicking that element is simpler, even if what happens once they're dissolved is equally complex, because at this point we're just talking about wood. Wood hardly needs to be barrel-shaped. Smaller tree fragments—staves, chips, pellets, even sawdust—can be steeped in a stainless-steel wine tank as you would tea in a mug. About 25 grams of oak chips per liter of wine in a tank provides the same oak surface area exposure that wine would see in a standard 225-liter barrel. From that surface seeps sensory-active compounds and phenolics that modify color and texture, far less expensively and with none of the various inconveniences of a barrel. The larger the surface area, the faster flavors diffuse, so using more chips (or what have you) also accelerates the process.

Oak's boldest—or maybe that's "vulgar"—expression is hard to miss in a brassy chardonnay, but most of oak's influences are more sensorily subtle and chemically complex. While food chemistry researchers continue to characterize oak components of newly appreciated sensory significance, the major players are easy to name. Wood cell walls are largely cellulose, relevant for wine only as a structural element that holds everything else together. Next, about 25 percent of wood solids are hemicelluloses, sugars that caramelize when heated to create sweet, toasty-roasty notes. A whole set of diverse polyphenols contribute spicy, nutty, smoky, or roasted aromas in varying proportions depending on the wood's provenance and the toasting it's seen.[7] Among them, lignin deserves special mention: it's the most abundant (20–25 percent of the wood's dry weight), chiefly responsible for vanilla flavors, and the source of artificial vanillin produced from wood waste.[8] Eugenol, another phenolic, is the major component of clove oil and forms in oak during the requisite air-drying step that precedes barrel construction. Whisky lactones, named for the coconut aromas they contribute to whisky, can contribute similar aromas to wine. Gallic acid (see the GALLIC ACID box) caps pigmented anthocyanin

---

GALLIC ACID

Gallic acid was initially identified in oak galls, infection-induced tree growths used in iron gall ink, Europe's most common writing and drawing medium until the nineteenth century. Gallic acid contributed to color stability in that context too. It came to pass that its innate corrosiveness wasn't compatible with new papermaking techniques. Unfortunately, the iron ions in these inks steal electrons from oxygen in just the same way as iron ions in wine do, ⊛ producing highly reactive oxygen radicals that destroy structural components of writing surfaces on their way to finding the electron they lack. As a result, valuable historic documents—writings by Galileo, drawings by Leonardo Da Vinci, musical scores by Bach, and so on—are inclined to fall apart without careful attention from conservators to neutralize acids and apply counteracting antioxidants.

---

[7] Barrel staves are always heated in the course of being bent into shape, but extra toasting is often applied solely for flavor. The staves employed in any given barrel will likely be exposed to different degrees of heat during construction, so one barrel will carry a symphony of sensory notes. That effect can be replicated with a mix of chips toasted to different degrees; uniformity isn't necessarily the goal.

[8] Much of the world's artificially produced vanillin has been made from wood waste. That's changing as wood pulping processes become more efficient and leave less waste behind. Among other ways to fill the gap, some vanillin is now being produced by *Saccharomyces cerevisiae* engineered with the genetic pathway necessary to produce that molecule.

molecules, adding to the color stability of barrel-aged reds. Oak-derived tannins, structurally distinct from the tannins derived from grapes, also modify color and texture.[9] All deplete over time; the more times a barrel is filled, the fewer soluble components remain. Reused several times, "neutral" barrels leached of their would-be-liquid influences effectively become their own form of micro-ox machine.

Soluble oak-derived compounds are sometimes described, disparagingly, as flavorings.[10] That's not entirely fair. Flavor is the entire point for some oak products—chips, staves, even liquid oak extracts—with names like "spice rack" or "sweet shoppe" and bold (or vulgar, as you will) caramelized hemicellulose-derived sensory notes. But when less caramelizing is involved, a complex array of phenolics are oak's principal contributions, and phenolic contributions go far beyond flavor as such.

Phenolics, a large and diverse class of molecules identified by their six-carbon rings, are the most common group of metabolic byproducts across plant life (see Figure 12.1). They include the tannins chiefly responsible for wine's astringency, anthocyanins chiefly responsible for its color, and a miscellany of more obscure molecules. Most matter to wine quality in one way or another. Many are antioxidants that mitigate the varied effects of oxygen on texture, color, and flavor. Resveratrol, a red-grape constituent sold as a nutrition supplement (for being an antioxidant, among other things ▲), is a phenolic. Oak hardwood is rich in phenolics not found in grapes that synergize with grape-derived counterparts to reconfigure their overall color and astringency effects.[11] In short, calling them "flavoring" doesn't do them justice.

---

[9]   García-Estévez et al., "Effect of the Type."

[10]   Oak is the only legal means of adding flavors, beyond those derived from grapes, to anything that will be sold as "wine." Yeast also add flavors to wine beyond those originally present in the grapes. However, strictly speaking, those contributions come from yeast metabolizing grape sugars, so even flavors derived from the physical breakdown of yeast cells themselves, and certainly the metabolic byproducts they throw off, can be described as derived from grapes.

[11]   Namely, ellagitannins and gallotannins, the latter of which is so called because can be digested to form gallic acid.

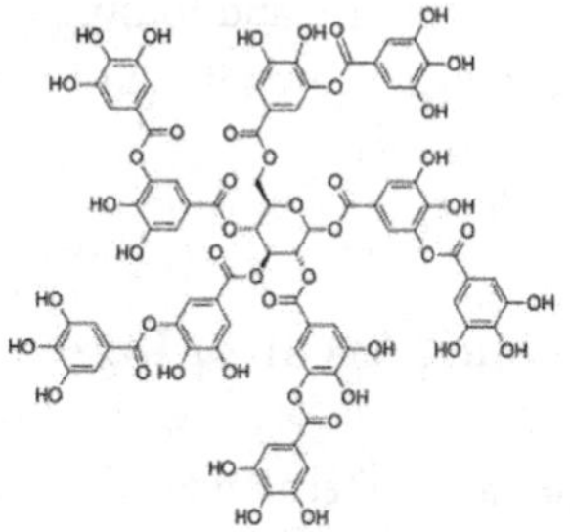

Figure 12.1 Phenolics have lots of charged side chains that tend to associate with each other and with proteins, 🦠 forming much larger complexes.

We can't even just talk about oak phenolics *adding* to the pool of grape-derived phenolics, because if this class of molecule does one thing, it's bind, react, and recombine with other members of its kind. How, and to what effect, constitutes one of the knottiest sets of problems across wine science, because phenolic reactions are a function of their respective proportions as well as their precise kinds, how much oxygen is around, and other elements of the overall wine environment. For example, tannins concatenate over time to form increasingly long molecular chains. In general, longer tannins feel less astringent than shorter ones; tannin unit for tannin unit, they present fewer opportunities to bind to salivary proteins. 🦠 Consequently, rough young reds smooth out with age.[12] However, grape-seed-derived tannins are shorter than grape-skin-derived tannins, and yet seed tannins feel more astringent. In experiments using standardized tannin-binding proteins, overall tannin concentration doesn't directly correlate with their "stickiness" at all.[13] And, as you might expect, tannin concentrations and ratios differ from stave to stave and

---

[12] Tannins will aggregate with proteins into such large molecular complexes that they become too heavy to remain suspended in wine. If the wine is in a tank or barrel, those aggregates fall to the bottom and can be removed—another reason why wines become less tannic with age. If the wine is in your mouth, those aggregates contribute to the rough feeling known as astringency.

[13] Revelette, Barak, and Kennedy, "High-Performance Liquid Chromatography."

barrel to barrel, with provenance and toasting, so predicting precise effects would be a difficult game even if their behavior was thoroughly mapped out—which it's not, yet.

Processing Agent, or Ingredient?

Combined, micro-ox and alternative-oak technologies decouple adding oxygen from adding oak-derived soluble molecules, and decouple both from the practical necessity of containing liquids. These opportunities to break with inconvenient, expensive barrels have been welcomed with open arms in many cellars (though others excoriate them as unwarranted interventions). 🌿 Broadly, the wine industry has made room for countless new processing agents. In contrast, industry regulators have been stingier about approving new *ingredients*.

The general rule is that all wine ingredients should either be derived from grapes or be typically found in them, such as sugar or acid. Oak barrels and sulfur dioxide are grandfathered-in exceptions. So are additives that *could* be considered ingredients, such as enzymes that kill bacteria or help release aromas, but which are instead classed as processing agents, maybe for the sake of consistency with the general rule.

So where do oak-derived products fall? New labeling regulations require rearticulating a detailed answer to that question. By 2023, wine and other alcoholic beverages sold in the European Union will be required to carry both calorie counts and ingredient lists, at least via a QR code through which smartphone users can find that information online. After decades of debate about whether alcoholic beverages should be exempt from labeling laws that apply to other food and drink, the industry agreed in 2018 that nutrition-y labeling is a good idea in principle. But as of the end of 2021, the details of what that means have yet to be finalized. The supranational Organisation Internationale de la Vigne et du Vin (OIV) is still finalizing the list of "additives used as processing aids" that won't require disclosure, in contrast with "additives during the winemaking process" that will. Barrels will unquestionably be on the first list, along with oak fragments that mimic barrels. But what about liquid oak extracts, or liquid or powdered tannin products *derived* from oak but not added through exposure to oak? The OIV has already decided that ingredients won't include the "controlled addition of natural substances during the process, when needed, to adjust and rebalance the composition of the product," including sugar and acids.[14] The rationale goes that wine is an "agricultural product" that relies on variable constituents; as such, additions are sometimes required to compensate for what any given year's grapes may lack. That logic seems weak, since other nuances of how labeling will be applied *rely* on wine varying from year to year and assume that grapes *aren't* consistent. A more internally coherent rationale may be that storied European wineries have long relied on sugar, acid, and sometimes other "corrections" but don't want to have to say so.

Where does that leave oak, or oxygen? Newfangled oak-derived tannin extracts could be considered compensation for tannins that grapes happen to lack. In contrast, traditional thoroughly toasted oak barrels look far more like flavor additions for the sake of a particular style. And yet the former is stirred into wine and left there, whereas the latter are physical objects brought in only temporarily. Meanwhile, oxygen may not feel like either an ingredient or a processing agent, but micro-ox now uses concentrated oxygen, which has to be added deliberately.

How any of these barrel-replacing . . . elements . . . is officially classified will probably be more about marketing than either science or tradition. Which, of course, is where we started, with butter-bomb chardonnays that win over restaurant patrons even though they're anachronistic, inconvenient, and crass by most authorities' standards. I admit: I like barrels (even if I don't much care for overtly oaky wines), not just because I enjoy how each one is different from the next, and not just because I appreciate fine woodworking, but because barrels are illogical and even a bit silly. If nothing else, oak is a superb reminder that neither logic nor physics and biochemistry hold an authoritative trump card over wine practice at the end of the day.

---

[14] Laaninen, "Alcohol Labelling."

13

Waste 🐌

On a global scale, the wine industry is a wholly insignificant player in humanity's race toward self-immolation by way of planetary heating. On a personal scale, plenty of wine people in decision-making roles are galvanized by environmental concerns. Marketing research suggests that you, as a wine drinker, probably aren't thinking about carbon footprints or sustainability claims when you decide which bottles to buy, and there's no reason you should; your individual drinking habits, no matter how virtuous, are insignificant to the global sustainability crisis. And yet marketing wine regions' various sustainability programs—because nearly everyone has one—has attracted academic and industry attention way out of proportion to the everyday consumer's inclination to care. Scale, it turns out, is a common theme in sustainability research.

I don't want to valorize baroque and obtuse wine sustainability certifications that, in addition to being obtuse, are sometimes more about image than substance. I *do* want to talk about how myriad ways of making knowledge contribute to the collective pursuit of delicious accompaniments to dinner. And so here I am, writing about winery waste, because it's an active scientific subject, and because waste is an inevitable and yet often overlooked component of bringing wine to our collective tables. Magicking loads of stinking, noisome grape leftovers into profitable products has become a widely appealing proposition, leading to novel grape-waste-derived products that range from skincare to anti-parasite treatments for honeybees.

Sticky Solids

Wineries are subject to homeostasis just like everything else. What goes in must come out. When truckfuls of grapes are loaded through their doors in the fall, we know that an equivalent amount of mass must reemerge.

The liquid fraction is relatively easy to account for, though the ocean of water used in wine production is the subject of its own sustainability conversation. The solid bits are stickier. Grape skins, seeds, and stems, collectively called pomace, are responsible for 20 to 30 percent of a harvest's total weight. They're just parts of plants, so they might seem benign, but pomace is regulated as a form of pollution for more than one good reason.

Like grapes themselves, grape residue is inhospitably acidic, around pH 3.5 or so. It's usually even more concentrated than whole grapes in tannins and other antimicrobial phenolics. Microbes therefore won't just jump in to casually decompose the stuff, as they will for your average pile of organic matter. Pomace is also sugar-rich. Sitting around, it stinks and attracts insects and larger pests. If permitted to seep into ponds and streams, pomace runoff will feed microbial overgrowth in waterways, using up dissolved oxygen and killing fish and aquatic plants—a process called eutrophication, often associated with fertilizer runoff from intensive agriculture. Disposing of pomace therefore requires care, but in the absence of better disposal solutions, it ends up in landfills or burned. Neither is good.

Waste into Gold?

A classic disposal solution is grappa, the Italian name for the peripatetic brandy made by soaking pomace in water and distilling the results. Converting solid waste into even more liquid might not seem helpful, but since pomace's high sugar and phenolic content is much of its trouble, extracting and making separate use of those components effectively neutralizes the toxic potential of what's left. Under the EU's strict provenance-based labeling laws, "grappa" can only be made in Italy, the culturally Italian portion of Switzerland, and the microstate of the Most Serene Republic of San Marino. All the same, you'll find similar products everywhere you find wine: marc in French-speaking places, orujo in Spanish-speaking places, tsipouro in Greece, and so on. (US-made pomace brandies often adopt the Italian name.) We're also beginning to see a renaissance for piquette, the mild beverage made by soaking and fermenting pomace without bothering to distill it afterward.

Until recently, the European Union mandated that grape pomace and lees (sediment collected from the bottom of aging vessels, largely consisting of dead yeast) be sent to distilleries, and distillers received government subsidies to perform the essential function of processing it.[1] But the products of distillation are in lower demand than they used to be, and distilleries are far from the most sustainable places in the world; they're often energy hogs, and they do expel nearly as much solid waste as they ingest, some of which ends up in landfills. Again, not great.

Conveniently, humans aren't the only traditional target consumer for pomace. Other critters will eat the solid parts. In the right neighborhood, you'll see pomace-fed pigs, turkeys, and other livestock, a classic small-scale solution that the animal nutrition folks have recently been bolstering by investigating the finer points of digestibility that matter at larger commercial scales. The limiting factor is that tannins and other phenolics make micronutrients less bioavailable, so too much pomace reduces an animal's ability to liberate vitamins and minerals from its diet on the whole. Compared to hay, pomace is also fairly nutrient-poor in the first place; it's mostly bulk. And animals are understandably disinterested in eating the stuff when it's anything but fresh—a serious limitation when all of the grapes in whatever half of the world you inhabit ripen in the span of two or three months. The upside is that livestock seem to benefit from an antioxidant-rich diet just as much as humans do, though whether they pass those benefits back to humans in milk and meat seems to depend on a lot of other contextual particulars.

Feeding pomace to microbes is difficult—again, compost doesn't just happen—but it's certainly not impossible. Adding lime to raise the pH to more microbe-friendly territory, plus some nice carbon-rich brown material (grapevine prunings, maybe), plus lots of turning, will eventually convert raw grape gunk into a beneficial soil amendment. Turning a compost pile incorporates oxygen, and up to a point, more oxygen accelerates decomposition. The city of Napa, California, which mandates that wineries, other businesses, and residents compost as of January 2022, has a two-acre concrete "active composting pad" underlaid with a network of pipes that inject oxygen into the pile from

---

[1] EC 479/2008. It has also been de rigueur for smaller wineries to ignore this mandate, since the cost of transporting their waste didn't match what they got out of the deal.

below through 3,300 nozzles. A covering of wood chips or mature compost retains heat, smells, and CO_2, altogether compressing what would otherwise be a six-to-ten-month process into less than a month of active time plus a month and a half of curing.[2] Working at scale—with a matching investment in infrastructure—makes all manner of things possible.

Other "waste valorization" options are also made economically viable by economies of scale, and by winemaking regions dense enough to support businesses that exist solely to strip value from what wineries throw away. Grape seeds, for example, are 10 to 20 percent oil. You may have personally participated in the circular wine economy by stir-frying something in grape seed oil, which has a high smoke point and a nutritionally favorable fatty acid composition. Defatted grape seeds can subsequently be pulverized into antioxidant-rich flour, the kind of ingredient that manufacturers of "functional" processed foods sometimes incorporate. And most of the tartaric acid on the market, which bakers will know as cream of tartar and an essential aid to whipping egg whites or making snickerdoodles, is stripped from winery waste.

Waste into More Gold?

As far as white powders go, tartaric acid is hardly high-value—a spice jar's worth is a few dollars at any grocery store—though the industry was still worth US $2.55 billion in 2020.[3] Converting winery waste into higher-value products is a work in progress. "Pomace extract" seems to protect mice against diabetes and heart disease. Both conditions develop through damage caused by highly reactive oxygen radicals, so concentrated antioxidants slow them down. Pomace is replete with antioxidants. However, developing new drugs requires massive upfront resource investments by a company that expects to make even more money in the long run, so less profitable potential innovations can easily languish on the R&D shelf. Thus far, only one grape-derived antioxidant, resveratrol (see the RESVERATROL box), has taken off as

---

[2] Goldstein, "Solid Waste."
[3] Research and Markets, "$2+ Billion Worldwide Tartaric Acid Industry."

RESVERATROL

Resveratrol is a phenol, one of a large family of molecules that also includes astringent tannins and colorful anthocyanins. 🛢 Like its chemical cousins, it functions as an antioxidant, truncating the chain of events that begins with oxygen losing an electron and ends with widespread damage to other molecules. As cardiovascular disease develops, this "oxidative stress" roughs up usually smooth blood vessel linings, creating sticky spots to which cholesterol adheres, narrowing and stiffening the blood vessel so that the heart has to work harder to pump blood through the body. Resveratrol seems to deflect this chain of events not only by securing oxygen radicals, but also by stimulating blood vessels to produce nitrous oxide, which prompts them to relax and consequently lowers one's blood pressure. Nitrous oxide also helpfully makes platelets less sticky and less inclined to glue together cholesterol in the first place.

a dietary supplement. It's commercially available as a supplement, but it still has clinical trials to pass before it can be officially recommended as a therapeutic agent with specific benefits.

Consumer products aren't the only way to make waste pay off. A winery's energy bill can be astronomical, so recycling waste into electricity is another aspirational goal—not least because it might deal with second-order sludge left over from other waste management strategies.[4] The simplest option is to burn dry solids to make heat. That relatively inefficient strategy can be improved with the right infrastructure, but for all but the largest operations—or cooperatives of smaller operations—investing in that infrastructure will cost more than it pays back.

---

[4] *Can* be, because energy-efficient designs are another avenue for trying to minimize the environmental impact of the wine industry, from gravity-flow wineries, where grapes fall from one step of the process to the next, to better insulation and passive solar designs to regulate temperature.

Pomace can be made to burn more efficiently by compressing it (ideally after valuable oils and such have been extracted) into pellets that burn like charcoal, given an outfit to produce the pellets. Pomace can also made into charcoal by heating it in the absence of oxygen, but it's now far more likely that anyone going to that trouble will make use of recent improvements on that ancient art that make more than just charcoal—and that require even more infrastructure. In contrast to the days-long traditional charcoaling process, contemporary "flash pyrolysis" can be close to instantaneous by applying heat at rates as high as 1,000°C per second. In the presence of oxygen, pomace would simply burn. Without oxygen, heating it to temperatures far higher than those at which it would typically combust (about 300°C for wood) precipitates a pile of reactions that end in liquid bio-oil and solid bio-char. Both can be used as fuel. Biochar can also be tilled into fields to support soil structure and capture carbon.

For larger winemaking operations that crush at least a thousand tons of grapes each vintage—that's about 750,000 standard 750 mL bottles—a pyrolysis unit and the equipment to use the bio-oil it generates might pay for itself.[5] Fewer than 3 percent of wineries in the United States make that much wine. For everyone else, a mobile pyrolysis unit on the back of a truck can be rolled up the driveway when needed, but may cost more to run than it pays in energy savings at the end of the day.

Finally, remember that troublesomely high sugar content? It makes pomace a potential feedstock for bioindustrial processes that convert sugar into useful industrial molecules via genetically engineered microbes. 🔊 Food flavorings, cleaning agents, carpet fibers, and a theoretically endless array of other things are now being fermented, with yeast and bacteria, in tanks that look just like those used to brew beer. In an ideal world, those production lines reclaim sources of carbon that would otherwise be wasted. The current world isn't ideal. Among other things, the range of practicable feedstocks for bioindustry remains narrow, and microbes are sometimes fed sugar from crops grown in lieu of people food or creature habitat. Pomace would rarely

---

5 Zhang et al., "Sustainable Options."

be the cheapest or easiest-to-transport waste to feed to microbes, but at some times of year in some regions, it should be—if someone is there to connect the supply chain.

... Or Just Getting Rid of the Stuff

That's the point. Whether any theoretical solution to waste is a tenable practical solution comes down to scale and economies. So we circle back to the central sustainability problem. Every last bit of a grape can be made useful in one way or another. But the vinous equivalent of nose-to-tail cooking takes money, time, effort, and space. Just like breaking down an animal, someone has to break down that grape. Usually multiple someones. Those relationships have to be sustained and managed—more work to be done. Literal tons of gunk need to be moved around.[6] And that's all in addition to processing the wastewater that flows downstream from the copious washing up essential to sanitary winemaking.

If you live in an area where single-stream recycling and compostable food and garden waste are picked up at your home alongside your trash, managing your household waste is a minor commitment. Everything goes into one of three bins. Someone else handles the rest. Otherwise, compensating for the lack of those services may mean sorting waste into six or seven bins, all of which need to be stored somewhere until you have time to drive them to a recycling center, all of which is work *you* have to do on top of maintaining a backyard compost pile—if, that is, you're privileged to have a backyard and to not have raccoons. At some point, that work becomes unsustainable. You don't, and can't be expected to, have time and space and mental energy to do everything on top of the everyday business of being alive in the twenty-first century. And trying to do one thing well may create new problems—all of that driving to and from a recycling center, for instance.

---

[6] Biodynamic wineries are an exception. Biodynamic principles envision the farm as a self-sustaining system whose waste should go back into sustaining it, so every bit of waste is—in theory—managed right there on the farm. Compost is big in biodynamics.

Sustainability is an institutional and infrastructural problem, not an individual one. In densely populated wine regions, like Napa, a truck can make the rounds to convey waste to centralized processing facilities. In regions where wineries aren't so tightly packed, wealthier operations that choose to commit the necessary resources might manage sensibly on their own. What happens in the middle is the difficult part. So here's the challenge: where does the cost of expanding infrastructure meet the wineries in the middle?

One boutique winery's waste management plan is a fruit fly amid the landscape of global heating, but may be very important indeed to its immediate neighbors. A regional wine industry's water reduction plan won't change widespread drought conditions, but may make a difference to one town with a failing reservoir. The only thing I, individually, can do about Big Oil and Big Gas is vote for politicians who want to eliminate their subsidies and hold them accountable for the mass genocide of humanity in the name of greed. At the same time, my (relatively resource-rich) neighbors and I can still choose to grow our own vegetables, reduce our reliance on packaged foods purveyed by mega-chain supermarkets, and participate in a culture shift toward building local infrastructure for good living, because those moves will help *us* live well.

14

Flavor ↷

Can someone tell you that what you're tasting is wrong? Any reasonable answer has to integrate both "no" and "yes." Flavor is integration through and through, from how flavorful molecules combine in solution to how human nervous systems integrate sensory inputs and how perceptions become socially sensible through language. And so flavor is simultaneously an idiosyncratic individual experience and a shared one—hardly a groundbreaking thing to say, but absolutely vital to how it's studied.

Even though scientific instrumentation can quantify sensory-active molecules, taste can't be separated from who's doing the tasting and in what context. The answer to "Who's doing the tasting?" may occasionally be a computer, but computers still aren't very sophisticated about reporting on wine flavor. Real live humans, as it happens, are still the best tool for that job—in part because wine is complex, and in part because human perception is complex.

Mechanical tasters replace humans in plenty of other settings. Per Wilbur Scoville's original 1912 method, a chile's heat was judged by a trained panel who ingested increasing dilutions of that chile in sugar water and spat out Scoville units on the basis of when its heat became imperceptible. That protocol has now been replaced by direct measurements of spicy chile capsacinoids via high-performance liquid chromatography. Machines never experience the burnout that humans suffer with repeat exposure to spicy foods, eliminate person-to-person variation, and ensure that commercial hot sauces remain consistent from bottle to bottle. But they're tasked with quantifying only one uniquely important group of molecules. Similarly standardized reporting on more complex flavors has been a much bigger challenge.

Analytical chemistry tools can separate components of a mixture, generate a "flavorome" analysis accounting for a theoretically unlimited number of molecules associated with smell and taste (the ones we know about, at least; see the GAS CHROMATOGRAPHY-OLFACTOMETRY box), and algorithmically align flavorome profiles

GAS CHROMATOGRAPHY-OLFACTOMETRY

Not all tasters are either human or machine. A few, like gas chromatography-olfactometry, are cyborg, leveraging the abilities of analytical machines to disentangle mixtures and the abilities of analytical humans to interpret sensory properties. Gas chromatography is a method for unmixing a mixture by flowing it through a tube packed with a material that interacts with different molecules to unique degrees. Components of the mixture emerge from the end of the tube at different times, according to how strongly they interact, and are recorded by a detector. In gas chromatography-olfactometry, two detectors are paired: an analytic tool like mass spectrophotometry that informs on molecular structure, and a person who sits with their nose at a sniff port and waits for odors to come down the line. Machines can't tell you what something smells like, so this strategy makes it possible to match molecular descriptions with sensory descriptions, and to run flavor-related experiments even when the specific molecular composition of the odor-active molecule isn't known.

with desired wine styles. What they can't yet do is integrate wine components that have yet to be characterized, identify how molecules synergize to yield an overall flavor impression, or account for how individual humans experience that flavor.

Machine-learning specialists have been advancing on that problem in the name of targeting consumer preferences. Classifying consumers on the basis of stereotyped preferences is a market research standby. Given contemporary computing, it's relatively trivial to measure a panel of parameters—acids, sugars, tannins, several hundred high-impact "signature aroma compounds," and so on—and then algorithmically arrive at a representative chemical composition for wines that Australian women aged eighteen to thirty-five tend to like, or that are preferred by British drinkers of New Zealand sauvignon blanc versus their Chinese counterparts. Such alignments can be made on the basis of what sells where. Given enough data about what kinds of wines *you* personally enjoy, they might even recommend which wines are most likely to appeal to you—your little market segment of one—much as music-streaming apps serve up new tracks on the basis of what you've already given a thumbs-up.

Like music algorithms, wine algorithms are subject to the anti-discovery problem: they can't recommend things that you haven't yet learned that you like. They also still can't predict the synergistic effect that a wine's components will have when stitched together in your glass, or the flavor that you'll experience when ingesting them; they work on the basis of chemical profiles and can only report on flavor when human tasters come in for additional analysis. To explain why, we need to rewind and review how flavor works for humans.

Smell and Taste

The human sensorineural apparatus remains the most sensitive tool available to assess wine flavor. We have to use the polysyllabic mouthful "sensorineural apparatus" because we're not just discussing the nose and mouth. The most significant contribution to the experience known as "flavor" comes from your brain. But first we smell, inhaling aromatic molecules that reach the nose before the solids or liquids from which they emanate reach the mouth.

The bank of cells residing on the roof of your nasal cavity hosts about 420 variations on the olfactory receptor theme. When activated in combination, they add up to more than the sum of their parts, discriminating thousands of aromas. Noses paint the picture; tongues report on nutritional relevance via molecules that matter to food quality but aren't volatile enough to reach noses. Most of us have probably experienced the disorientation that comes with a head cold or some other (temporary, I hope) anosmia and finding that our usual ability to make sense of food is utterly destroyed.[1]

---

[1] If you haven't, you can temporarily produce it by trying to enjoy a meal while wearing one of those little plastic nose clips sold for use while swimming—less pinchy than a clothespin, but enough to keep external air from reaching your olfactory receptors.

Taste receptors only pick up sweet, salty, sour, umami, and bitter. However, the old idea that the tongue is compartmentalized into distinct regions specialized for only one of the five flavors is wrong. Taste buds are distributed across the tongue, roof of the mouth, and back of the throat. Each bud includes multiple kinds of receptors, so every flavor registers everywhere. Where we see regionality is in how those taste sensations make their way to the brain. Three distinct nerves are responsible for conveying them from the tongue, roof of the mouth, and throat. That redundancy means that minor mouth injuries can't easily wipe out our sense of taste; ageusia, or total loss of taste, typically happens through damage to the nerves themselves, not the receptors. By selectively anesthetizing those nerves, neuroscientists have confirmed which sensations aren't tastes at all, such as astringency, which people can *feel* even when they can't taste.[2]

The big five of taste types might give the impression that mouths are rudimentary in what they add to flavor, but we can't ignore that they register contributions that aren't taste. Mouths have mechanical receptors that report physical sensations such as the viscosity of a liquid, the tiny explosions of carbonation, or the literal roughness of tannins. 🛢 A different set of receptors that trip the trigeminal nerve receive sensations that seem physical but trigger a technically distinct kind of chemical sensation, including astringency, the prickle of dissolved carbon dioxide, or the similar but worse prickle of getting powdered potassium metabisulfite in your face. 🌐 The heat of chiles and the coolness of mint are also both chemical sensations. Oral mucosal surfaces have temperature receptors and pain receptors too, and all of these varied inputs participate in our conscious experience of flavor. Mouths may seem like the most mundane thing ever, but even in the decade since I sat in food chemistry class, the list of what we know they do has continued to grow.

All the same, smelling is what provides nuance. Smelling happens through two paths: orthonasally, while the sniffable thing is outside your mouth; and retronasally, after ingesting the aromatic object, when aromas travel up to your nose from the back of your throat. In both cases, noses exclusively register volatile molecules—those connected

---

[2] Schöbel et al., "Astringency Is a Trigeminal Sensation."

only tenuously to their solid or liquid surrounds such that individual molecules make their way into the air. The physical constraint of volatility is why you can't sniff the difference between a cup of salt water and a cup of sugar water; both sugar and salt bind too tightly to water to let go and fly upward. (If you're thinking that you can detect a difference: the aroma you perceive above a cup of salt water is iodine or other dissolved minerals present in sea salt but not in sugar, rather than the sodium chloride itself.)

Putting something in your mouth itself can alter its aroma. Salivary enzymes initiate digestion even before you swallow by breaking bonds among glucose molecules. Their activity explains why starches will begin tasting sweet if you chew them long enough.[3] In addition, those enzymes can also release volatiles by separating them from glucose molecules that otherwise keep them from evaporating. Those newly released aroma molecules can then make their way up the retronasal path and trigger sensations that weren't possible to sense before sipping. Microbiota resident in the mouth also get in on early digestion action, so variability in our oral microbiomes contributes to interindividual differences in flavor perception. So does salivary pH, which changes throughout the day in response to stress, among other things.

Person-to-person variation in all of these factors means that taste and aroma change in your mouth in ways that may differ from how they change in mine. Of course, individuals also vary in the sensory receptors they possess on account of genetics, age, sickness, exposures to nicotine or other drugs, and so on.[4] But because of saliva, we may be grappling with differences in what molecules even exist for those receptors to receive.

---

[3] Salivary enzymes are also why some ancient ferments, such as kuchikamizake ("mouth-chewed sake"), a hypertraditional ceremonial sake, begin with chewing grains and spitting them out. Salivary enzymes take the place of malting to convert starch to sugar.

[4] Many genetic differences in sensory capacity are minor, but some are huge. A recent study conducted in Pennsylvania found that about 40 percent of the casual consumers they surveyed were unable to detect rotundone, the compound that conveys peppery qualities to some shiraz (Gaby et al., "Individual Differences"). It's widely suspected that about a third of people working in the wine industry can't perceive 2-acetylpyrroline, which contributes to "mousiness," described as the sense that a wine has been aged inside a rodent cage by those who are sensitive to it. I wouldn't know, since I'm not.

Smell, and Taste, and Thinking

Brains add another layer of potential difference. Brains produce flavor sensations by integrating signals from sensory organs with other contextual cues: memory, price, physical and social surroundings, and so on. If your sense is that wine tastes better when you think it's expensive, for example, research will back you up. In neuroscientific terms, price is an "external cue" that modifies your physiological experience of something, even though that cue is "external" to the thing itself.

Roughly thirty brain regions are active during wine tasting. Among them are a small cluster of neural neighborhoods—between your eyebrows, about an inch deep—that also activate in response to receiving money and during the placebo effect.[5] That cluster lights up when a casual wine drinker, tasting a €12 wine, experiences it as more pleasant when they believe that it costs €18 than when they believe that it costs €6.[6] You could say that (ostensible) price is a kind of placebo, tricking you and your brain into thinking that you're receiving something that you're not—at least not in strictly external terms. The perception is plenty real.

Charles Spence's Crossmodal Sensory Lab at the University of Oxford has defined the field of more-than-the-wine-itself sensory studies, examining how music, color, lighting, and myriad other external cues shape what tasters perceive. What they call "sonic seasoning" describes how, for example, red wines taste fruitier in the company of "sweet" music rich in bells and piano than when tasters are subjected to piercing, piccolo-heavy "sour" music. Tasters rate crisp white wines as more enjoyable when they're listening to flute music with a lively melody than to slower, somber string music full of harmonic dissonance; a dense red Bordeaux, in contrast, rates more highly in the latter sonic environment. The lab has found that expert wine professionals aren't immune to these effects; they've conducted the "sweet" and "sour" music test with multiple audiences, including attendees at a technical industry conference, and the same patterns hold.[7]

---

[5] The ventromedial prefrontal cortex, the ventral striatum, and the anterior cingulate cortex, more precisely.

[6] Schmidt et al., "How Context Alters Value."

[7] Wang and Spence, "Assessing the Influence." The 2016 International Cool Climate Wine Symposium, to be precise.

Obviously, a main point of cross-modal research is to help wine businesses design environments that help them sell more wine. But there's another point. Sensory studies often investigate simple phenomena so that effects of individual sensory inputs are easier to isolate. Those studies will never identify cross-modal effects. Spence argues that complex phenomena such as wine are useful research tools precisely because they involve the kinds of complex sensory interactions relevant to "real life" outside the lab.

Recognizing that flavor is a product of your own head doesn't mean that tasters simply make stuff up, or that no one can ever agree about a wine's characteristics. Far from it. Flavor is a social construct in how we handle it as a shared experience. Just as placebos can deliver real relief to people who believe that they've taken a painkiller, you "really" may taste blackcurrant in a wine because someone suggested that you should. ⒜ Then you may learn to associate those three syllables with what you perceive when everyone around you is talking about blackcurrant, thereby incorporating a sensation called "blackcurrant" into your own sensory environment. Learning to taste blackcurrant is simultaneously a matter of the power of suggestion and the power of social learning.

Social learning is about developing shared environments. "Environment" translates as *Umwelt* in German. But *Umwelt*, specifically, is a technical term, coined by a biologist and theorist named Jacob von Uexküll in the 1930s to describe the personal world that every creature inhabits (see Figure 14.1). His proposition was that every creature lives in its own world, composed solely of the things it can sense and the things it can act on; there isn't just one universal world, but innumerable worlds inhabited by innumerable living things. For example, UV light is only part of my *Umwelt* when I'm sunburnt. For a bee, UV light colors everything. I don't seem to be sensitive to the tetrahydropyradines associated with "mousy" characteristics in some wines, so mousiness isn't part of my *Umwelt*, but the pepperiness of rotundone is. For others, rotundone only exists as an unseeable, untasteable, unsniffable thing other people talk about.

When an aspect of your *Umwelt* is constrained by the basic constitution of your physiology—your sensitivity to rotundone, for instance—no amount of training will change it. However, humans (and other creatures) can expand their *Umwelten* in many respects, via what von

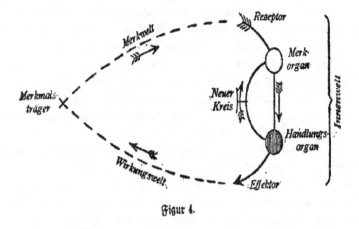

Figur 4.

Figure 14.1 Von Uexküll's "early scheme for a circular feedback cycle" of sensing and effecting, through which creatures constitute their *Umwelt* or individual environment, from *Theoretische Biologie* (1920). Image in the public domain.

Uexküll called "prostheses." That word may bring to mind robotic limbs or knee replacements, but in this context, we're talking about anything that conveys an extended capacity to sense or affect. I can't ordinarily see UV light like a bee, but I can augment myself with virtual reality applications that will (try to) mimic that ability for begoggled humans. If you can't taste rotundone, you can employ the prosthesis of a mass spectrophotometer to detect its presence in a solution. If you have no idea what blackcurrant tastes like, you can employ a friend who *does* know as a prosthesis to help you recognize when to talk about blackcurrant at a wine tasting.

Analytical scientific equipment—microscopes and mass spectrometers and the like—is a set of prostheses that expands the world of human perception. Education is another. Both are most useful when they're shared by a community, to increase overlaps among our shared environments, so that we can do more together.[8] That's what wine education is about: making

---

[8] That's true for the educational prostheses you might adopt to develop shared working worlds with fellow humans. Humans also educate themselves to work with other-than-human creatures: dogs and horses, obviously, but also wine yeast, which communicate in a multisensory language that humans can learn to understand.

space to work (and play, and enjoy) in a community dedicated to developing shared perceptions of something that we all think is worth learning to perceive in detail.

Smell, and Taste, and Thinking, and Training

The staff in detail-oriented restaurants rely on developing shared perceptions in shared environments to coordinate multiple cooks around a singular idea of how a particular dish should taste. Even though everyone's lived experience and physiological capacity to taste are unique and can't possibly be made the same, a group can learn to associate individual, idiosyncratic experiences with shared descriptors, and so generate an "objective" standard for the ideal flavor of a sauce. In that context, what a newbie *thinks* they taste will almost certainly be off-target until they learn to taste with the group. Individual experience is irrelevant, let alone personal preference; what matters is that everyone can reliably recognize and reproduce the sauce.

The same kind of social norming-learning-qua-peer-pressure is behind trained tasting panels used for research purposes.[9] When research requires ascertaining whether one yeast produces significantly different flavors in chardonnay compared to another, or whether a viticultural treatment decreases vegetal flavors in syrah, ten or twelve people need to taste as though they're just one person—one person who's somehow able to account for a range of interpersonal physiological variation. One superskilled taster won't do precisely because one person can't account for that range. Also, the scientific principle of replication says that drawing conclusions from one measurement is risky, because it could be off by a bit or even an outright mistake, while averaging several measurements provides much more security.

---

[9] Similar trained tasting panels are also used for institutional wine assessments, such as the "quality" panels that must approve Australian wines before they can be exported, or that confirm where a riesling should land on the International Riesling Federation's sweetness scale.

Replication is what's wanted from a trained panel. What's absolutely unwanted is a conglomeration of individual opinions about what constitutes vegetalness in syrah. To avoid the latter and achieve the former, panelists need to norm themselves to a singular, shared definition of the descriptor (see the ASPARAGUS box). You've been calling that mouth-feeling "furry" all your life? Nope, now it's "rough." You think this wine smells like the inside of your garbage can? The group is calling that "petrol," and you'd better learn to call it petrol along with everyone else or be unsurprised when you're asked to leave.

Becoming proficient at the team sport of trained tasting can be quite silly, and quite boring. The "reference standard" for vegetalness may be canned asparagus. To train on it, panelists may sample a prosaic syrah spiked with liquid from a can of asparagus. Then, like a machine-learning algorithm working out what its programmer wants it to recognize, the panel will work through a training set, discussing each wine and coming to consensus about how vegetal it is. Over. And over.

Once everyone has normed to the same terminology with the group, each taster will taste and rate a few wines of established vegetalness individually, on several separate occasions, so that the panelmaster/researcher can assess whether everyone is operating as a reliable analytical instrument individually and as a group. Only then, and only if everyone is in reasonable agreement on the training wines, will panelists

ASPARAGUS

Does asparagus taste the same to me as it does to you? Do we see the color red the same way? Whether or not everyone on a trained tasting panel internally experiences the same sensation in response to a shared stimulus isn't a scientific question that sensory scientists or even neurophysiologists can answer. That's a metaphysical question for phenomenologists, the guild of philosophers who concern themselves with what it means to consciously experience things. The answer doesn't matter for wine research purposes. So long as a team can learn to become a reliable collective analytical instrument, individual internal experiences are just as irrelevant as individual opinions.

individually rate the experimental wines that justify the whole exercise. Again, each person will taste and rate individually, generating ten or so individual measurements of the same thing.

If, in compiling everyone's results, the panelmaster finds that one team member's performance is inconsistent, with ratings that differ unpredictably from the rest of the group *or* from what they said about the very same wine last week, that member may be kicked out as an intractable rogue. Panelists earn their place by demonstrating reasonable sensory acuity and an ability to describe what they're tasting, along with plenty of perseverance. Whether they look like target consumers in terms of demographics or taste preferences is unimportant, because their reason for existing is to be a cyborg—a scientific tool with human sensitivities.

Trained panels are therefore no help at all in evaluating what consumers will think of a wine. All they can say about likeability is how a wine rates in qualities that *tend to associate* with likeability: lack of vegetalness in a syrah, intensity of passionfruit aroma in a sauvignon blanc, "dark fruit" character in a red blend, and so on. Assessing what consumers will think requires the inverse of a trained panel—an untrained panel of people who resemble a wine's target market. Though, in a pinch, underemployed graduate students have been known to do.

Even though they're essentially taste-and-tells, untrained sensory analyses are also completely unlike tasting under ordinary circumstances. First, the exercise typically occurs in a booth, with dividers separating you from the person in the next seat over so that you can't influence each other.[10] A little pass-through window lets a research assistant deliver and retrieve tasting trays from the adjoining room. Second, the researcher will have taken pains to remove potential distracting or confounding factors. If you're rating flavor intensity, the booth may be lit with red light so that the wine's color can't affect your judgment. If you're supposed to evaluate the sweetness of a cherry, you may wear a nose clip so that you aren't influenced by its overall flavor. Third, steps will be taken to reduce variability. Wineglasses will be covered so that aromas don't evaporate, allaying the possibility

[10] The International Organization for Standardization, responsible for ISO standards, even has ISO standards for the physical parameters of a standardized tasting booth.

that the length of the interval between pouring and tasting affects the study. Water and an unsalted cracker are usually available for palate cleansing between bites. Indeed, you may be required to sip water and munch a cracker to cleanse your palate between tastes. Tastes and the intervening rest-and-recovery periods may be timed.

All of this standardization enables reproducible results, but means that even untrained sensory studies are rubbish at reporting how you'll feel about a wine when you open it for a dinner party or what you'll think when you try it in a tasting room. Each of these is its own tasting context, as is a research exercise. Asking tasters to indicate how much they like a wine (a "hedonic assessment," in sensory science terms) will tell you how pleasant they find it on its own in a white-walled booth. An assessment over dinner will integrate a different set of multisensory inputs and external cues. Context matters.

Smell, and Taste, and Thinking, and Training, and Language

Trained panels rely on developing shared language. In the interest of bypassing language altogether, some neuroscientists have asked people to taste wine inside a magnetic resonance imaging machine—that is, an MRI. The idea is that imaging brains during the act itself should evaluate what tasters experience without fussing about whether one person's "a bit abrasive" is another's "unbearably tannic." It's the only way to see what bits of brain activate during wine tasting. And it might point to elements of experience that brains register but that people struggle to put into words, or that don't break through to consciousness at all.

An MRI scan requires the subject to lie extremely still. The "M" part—the tube-shaped magnet inside which the person being imaged lies—is powerful enough to compel your body's protons to align with its field. Then comes the "resonance" part, when the machine emits a radiofrequency pulse that spins those protons out of equilibrium, forcing them to work to find their way back to alignment with the magnetic field. Protons reequilibrate at different rates. The energy they release on their way back can be converted into an image, with contrast between one kind of tissue and the next provided courtesy of those different equilibration rates. As for that last "imaging" part, if you wiggle, so does the picture, and that's no good.

You can see the issue here. Wine tasting usually involves moving. Combining the two requires a setup that allows someone to sip while otherwise remaining perfectly immobile. In practice, that looks like a syringe squirting wine into your mouth while you lie flat on your back, arms held at your sides, inside a giant whirring magnet. A realistic tasting scenario it is not. It is highly controlled, however, which makes drawing conclusions about experimental variables far easier.

Researchers at the Basque Center of Cognition, Brain and Language used functional MRI (fMRI), as scanning people while they do something more than just lying down is termed, to compare how brains respond to wines that tasters don't consciously describe as different.[11] Their choice of brains, and thus tasters, is significant. The researchers wanted untrained reactions, not the carefully honed judgments of experts. So they asked twenty-one once-in-a-while wine drinkers to sip a 13–13.5 percent ABV dry red wine and a similar but slightly more alcoholic (14.5–15 percent ABV) vintage.[12] "Sip," remember, is a bit of a euphemism; tasters had small increments of either wine or a tasteless rinsing solution "delivered" into their otherwise still mouths while they were lying prone, with imminent liquid delivery heralded by an audible warning tone. Dispensing several sets of similar wines mitigated the concern that the study might learn about several wines in particular instead of alcohol differences in general, and tasters confirmed that they didn't detect much difference among the pairs when sipping from a glass outside the MRI.

Of the thirty brain regions that registered activity in response to wine but not the tasteless rinsing solution, only two responded differently to lower versus higher alcohols: the right insula, an inner fold

---

[11] Frost et al., "What Can the Brain."

[12] The researchers invited twenty-six volunteers to participate in the first instance, but five didn't make it through to the experiment. Three had to be eliminated because their baseline readings were too variable from measurement to measurement. Two were excluded because of unspecified unease with some part of the test. Losing a few subjects between recruitment and data collection is a routine part of conducting a sensory study, for which the number of initial recruits typically accounts.

of gray matter beneath your right temple, and the right cerebellum, at the lower back of the brain. Both were *more* active in response to *less* alcohol. The neuroscientists conducting the study expected the opposite. Common sense might suggest that more alcohol is more intense and therefore more stimulating. Instead, the study found that alcohol may dampen rather than heighten a wine's thought-provokingness, at least if "thought-provoking" means brain-activating. In addition, brains seemed to register responses more subtle than what tasters can describe.

Neurophysiological studies suggest that education—gaining the language, developing the prostheses—isn't about changing what our sensorineural apparatus can *perceive* as such, but about learning what to *notice*. Humans can perceive differences that they lack language to describe, and learning words to describe a sensation doesn't seem to alter whether or not we can detect it. Among other things, that inference comes from research that asks untrained tasters to identify which one of three wines (or cherries, or what have you) is different from the other two, without needing to say anything about how the standout stands out.[13]

Instead, neurologically, training looks like learning to direct one's attention. MRI-based inquiries have found that sommeliers apply analytic parts of their brain more intensively while tasting wine, while the brains of neophytes look like they're sipping sugar water.[14] The experts also know to sustain their attention through the post-swallowing, retronasal phase of tasting, while the plebes mostly think about wine while it's in their mouths.

Wine-describing norms have grown up around a European lexicon of sensory memories: blackcurrant, red fruits, elderflower, tobacco, leather, barnyard, sticking plaster, and so on. The eminently British Jancis Robinson's articulation of the signature note of sauvignon blanc as "cat piss on a gooseberry bush" is a sharp example; gooseberries are a British standby but hardly common everywhere. British wine experts seem to have a flair for this kind of thing. Oz Clarke—an unmistakably

---

[13] This is a formal sensory science method called a triangle test, useful when you want to ascertain whether two things are perceptibly different to real humans.

[14] Castriota-Scanderbeg et al., "The Appreciation of Wine."

English, cricket-playing, tea-drinking Oxford graduate—will go on at some length about "English hedgerow" if an aromatic English white wine gives him the opportunity.

The abundance of international students of wine who have been required to learn such references doesn't make those references universal. With growth in the Chinese wine market have come efforts to equip Chinese drinkers with more culturally appropriate aids, including a lexicon of Chinese wine aroma references. Those efforts have to consider what kinds of shared *Umwelten* are good to build within and across cultures, what kinds of aromas are notable, and which are useful. For example, Australian sensory scientists working with Chinese tasters report that people who buy imported wine in Beijing and Shanghai have highly varied tastes, like people who buy wine everywhere.[15] But the largest fraction prefer reds with robust fermented bean curd and white pepper notes, and are turned off by aromas of dried longan, waxberry, and pickled cucumber. In contrast, the preferences of the most wine-educated and wine-confident consumers in this study couldn't be defined in terms of flavor preferences at all.

That's telling. No matter what culture you grow up in, if wine is relatively unfamiliar to you, you may seek out wines characterized by a particular flavor you enjoy. You may not yet have learned how to *notice* anything else. With training, you may instead seek a sense of balance or harmony or energy that isn't directly associated with longan or gooseberry or any other individual flavor. Language changes how we experience wine, but not necessarily in obvious ways.

So What?

Being able to directly assess what someone thinks of a wine, without the troublesome interference of language, might seem like a winemaker's or wine marketer's dream come true. I don't think that it is. In prioritizing some ideal of objective, unbiased standards, it can

[15] Williamson et al., "More Hawthorn."

be easy to lose track of what we might be trying to achieve. What do patterns of brain activation say about flavor as an experience?

Something. Not everything. Neuroscientists have identified "reward centers," "pleasure centers," and "valuation centers." They can watch people associate experiences with memories and emotions when they route a sensory input through the amygdala, an almond-shaped bit of brain associated with memory and emotion. But even if conclusive evidence demonstrates that increasing alcohol concentration dials down brain activity, that hardly means that lower-alcohol wines are better. We can't say that more brain activation is good, or delicious, or desirable, only that it's *more*. Watching brain regions light up says very little about the experience of drinking, just as tasting notes that advertise hints of black pepper and blueberries say very little about how you might feel about the wine attached to those descriptors. At the end of the day, scientific research that wants to know about *perception* still has to rely on asking people questions, with all of the integration—of experience and language, of individual perception and social environment, of sensorineural activation and context—that entails.

15
Health 🍷

You probably aren't allergic to wine. I'm not just saying that because you're reading a book about wine. Decades of research on the subject says that wine allergies are extremely rare. On the other hand, you might not glean the same impression from surveying the room at your local watering hole. Self-reported wine-intolerance symptoms are legion. Are people who claim to be sensitive just making up their symptoms? That accusation would hardly be fair. It's also not scientifically reasonable. True allergies are rare, but "allergies" aren't the only way to have a bad reaction to wine—though some of those ways have more scientific support than others.

Allergies and Sensitivities?

An allergy, definitionally, is an immune response. Generally, the allergic person produces antibodies that bind to some molecular aspect of the thing to which they're allergic, setting off a chain of immunological events. The manifestation of those events might be fairly minor—a rash, say—but because immune responses can travel the length and breadth of the bloodstream, they might also include life-threatening anaphylactic shock. While allergies to anything are possible, some are more likely than others. Grapes are far from the top of the list. Indeed, when a group of German physicians encountered a patient who mounted an immune response to grapes fresh and cooked, raisins and wine alike, it was an exceptional case worth writing up in an allergy journal.[1]

Sensitivities, related to any number of other-than-antibody-mediated mechanisms, are categorically not allergies and are rarely

---

[1] Schad et al., "Wine Anaphylaxis."

such a big deal. For example, I'm among a subset of people with a relatively mild sensitivity to biogenic amines, molecules produced by some yeast and bacteria that resemble human hormones and that can leave me red-faced and headachy after just one glass—uncomfortable, sometimes embarrassing, far from life-threatening. In another familiar example, a substantial fraction of Asian people experience acute alcohol sensitivity because they don't metabolize alcohol itself well. For them, the flushing and general discomfort that set in shortly after imbibing can be blamed on a genetically encoded difference in one of two essential ethanol-decomposing enzymes.

If you're intolerant of alcohol, your sensitivity to wine probably doesn't occasion much mystery in your life. If you're intolerant of a substance that *sometimes* occurs in wine because it's *sometimes* used as a processing agent, the origin of your discomfort may be less obvious. We're talking about the usual suspects. Grapes are always gluten-free, but trace amounts of gluten can theoretically travel into wine from a wheat flour paste sometimes (increasingly rarely, as you might expect) used to seal fissures in oak barrels—potentially enough to trouble someone with Crohn's disease or a profound allergy, though not those who avoid gluten for lifestyle reasons.

Gluten-based products are also among a fairly long list of fining agents used to clarify many wines of suspended particulate matter. So are egg whites, milk products, gelatin, and a gelatin-like substance derived from fish bladders (yes, fish bladders) called isinglass.[2] (So is bentonite, a finely milled clay, though I've yet to hear of anyone being allergic to that.) Plant-derived agents are slowly replacing some of these, in part to assuage potential allergy concerns, in part to allay anxieties on the part of vegan and vegetarian drinkers, and in part because they're sometimes technical improvements over the old way. Regardless, fining agents are thoroughly removed before bottling; indeed, their whole point is to attach to other suspended solids, forming clumps large enough to either fall to the bottom or float to the top of a tank, where they're easily eliminated. All of that happens before the more extreme filtration steps

---

[2] Pliny the Elder, the first-century AD Roman classifier of many things, records that "fish-glue" was in use in his day—much the same as what's used now according to his description. Royle, *On the Production of Isinglass.*

to which many wines are subject, involving membranes and pad filters akin to ones you might have in your home air or water purifier, and which constitute another level of assurance that you're not getting a side of fish bladders with your Barbera. The result is that only oenophiles with severe allergies triggered by near-imperceptible molecular traces have cause for concern, and those concerned on moral principle.

In theory, all the careful winemaking in the world cannot absolutely guarantee that trace proteins from fining agents won't be left behind. Neither can official allergy-detection protocols of the kind government authorities use to adjudicate labeling laws. Newer methods have been finding fining-agent residues that official methods miss because the updated alternatives have been designed specifically with wine in mind—and wine, it turns out, is an especially tricky mixture to parse.

The typical gold standard for allergen detection is the enzyme-linked immunosorbent assay, or ELISA. Antibodies to the target molecule are affixed to the bottom of little compartments in a plastic tray. A diluted version of the food being tested is dispensed into the compartments and then washed away, so that only molecules specifically attached to those immobilized antibodies remain. A color-change-based readout visualizes where antibodies have found their target. ELISAs are incredibly simple to perform; with the right equipment, you could run one in your kitchen while you made a cup of tea. ELISAs also aren't perfect. They won't consistently detect traces of egg or gelatin proteins in wine because wine is also full of large, complex molecules (phenolics, notably) that physically impede antibody binding.[3]

More sensitivity can be achieved with multistep methods that mechanically separate a wine's components before trying to detect allergens. Those new methods raise a question: did the old methods need to be improved? In other words, just how many molecules of a potential allergen must be left behind to count as a contaminant? Since 2012, European Union legislation has mandated labeling for products containing more than 0.25 milligrams per liter of proteins from milk, wheat, or eggs. In the United States, the Food and Drug Administration links labeling to positive ELISA results, which can be significantly

---

[3] Dal Bello et al., "Multi-Target."

more sensitive than the EU threshold under ideal conditions. (Wine, remember, is not an ideal condition.) In theory, both rules allow the possibility of wine causing a problem for someone with the kind of allergies that require preparing all of one's own food and carrying an EpiPen. That said, the biomedical literature is hardly brimming with examples of allergy sufferers suffering this problem; on the contrary, it's hard to find any examples at all.

Wine headaches are another matter entirely. While numbers aren't easily pinned down on such things, some reports suggest that as many as a quarter to a third of drinkers find that red wine leaves them with a headache, even after a modest glass or two. The experience is so common that the International Headache Society (no, I didn't know that existed before either) has not one but two official classifications: immediate alcohol-induced headache for those occurring within three hours, and a delayed version for those occurring five to twelve hours after drinking.

Official classifications or no, medical establishments have sometimes been disparaging about wine headaches, maybe because they've yet to explain them. One potential culprit is acetic acid, the output of the two-enzyme chain that digests ethanol.[4] At least, it appears that you can give a rat a headache by feeding it rather a lot of acetic acid. Then again, acetic acid is just vinegar, and vinegar doesn't seem to cause most headache sufferers similar troubles.[5]

The story is similar with every other specific trigger that's been auditioned as the source of the problem: every time some poor molecule gets worked up as the putative headache instigator, the research doesn't quite add up. Histamine, tyramine, and phenylethylamine are all known to provoke headaches when found in cheese, cured meats, and other fermented products. Wine contains all three. However, studies that try to provoke or avert wine headaches by adding or subtracting those substances haven't reproduced meaningful patterns. Sulfur dioxide falls into the same category. More SO_2 tends to be used

---

[4] Krymchantowski and Jevoux, "Wine and Headache."
[5] Lacto-fermented pickles, made by salting produce and letting lactic acid bacteria do the souring, do show up on lists of foods that frequent migraine sufferers should avoid. The concern in these cases isn't the acid that bacteria produce but histamines and such that they may generate as byproducts.

in white wines than in reds, which doesn't fit the pattern of red wines causing worse headaches; moreover, reports of wine headaches haven't fallen off with trends to use less SO$_2$. 🅢🅞₂ And organic and other no-sulfur-added wines still give people headaches too.

A more likely set of suspects is flavonoids, a group of phenolics that give red wines color and texture. That lines up nicely with the difference migraineurs report between reds and whites. The idea of flavonoids as a trigger also aligns with tomatoes and chocolate being common headache inducers, since they're also flavonoid-rich. Then again, some flavonoid-rich plants, such as moringa, are traditional headache remedies.

At the end of the day, the problem may be that individuals who experience headaches have different triggers, so studies struggle to pick up on patterns. The resulting lack of scientific clarity doesn't mean that your headaches (or the headaches of your friend who's reluctant to share a bottle with you) don't exist or that you're making them up. It does mean that if you understand your headache triggers, you may have a better idea of how to avoid them than anyone else does.

Toxins and Contaminants?

But this chapter is about health, and health isn't just about immediate reactions. You might, instead or in addition, be concerned about long-term effects from what you drink. Perhaps you're thinking about your liver—reasonable. Perhaps you hope that your heart is thanking you for your daily glass of red—also reasonable. Before we get to those kinds of long-term effects, we need to deal with the unreasonable worry that your wine habit might be poisoning you with hidden toxins or additives. When I call them unreasonable, I mean that those fears aren't grounded in evidence, not that folks don't have reasons for thinking them. Regrettably, those reasons have been planted by a "clean wine" industry preying on people's fears to push their own purportedly safer products.

A genealogy of the clean wine fad would need to account for the far wider "clean" phenomenon (clean eating, clean skincare products, and so on). That's not my goal here. Instead, I want to make it clear that the main role of science in "clean" wine brands involves debunking their claims. But to do that, I need to point out that some of the foundations for the movement are sensible, while others masquerade as evidence-based while being anything but. In the first category is the sensible point that wine packaging hasn't historically carried nutrition information. That's legitimately troublesome for any number of reasons. Also in the first category, the rules about what's permitted and what's prohibited in wine are arcane and obscure, so consumers can't make easy sense of what goes into what they're drinking. In the second category, of faked science, are illogical outcries such as a wine-arsenic scare that attracted too much attention a few years before clean wine hit the scene, and that has achieved regrettable tenacity with some media outlets since then.

In 2015, an entrepreneur named Kevin Hicks decided to sell a service, BeverageGrades, to warn consumers about which wines were highest and lowest in arsenic and other toxins, calories, sugar, and carbohydrates. That service also promised recommendations about which wines were most "healthy," though what "healthy" meant was suspiciously unspecified. His marketing plan included a lawsuit against wineries that BeverageGrades had identified as producing wines that contained the most arsenic, which created exactly the effect that one might expect—entirely failing as a lawsuit but generating plenty of press.[6]

Wine does contain arsenic. So does, your drinking water and many of the foods you eat. Arsenic naturally occurs in small amounts in soil, air, and water. It's also a common soil contaminant, thanks to the legacy of arsenic-based pesticides, insecticides, and other "-cides" widely used in agriculture through the mid-twentieth century. Arsenic is still used in wood preservation, and it sticks around in soil rather than being broken down or washed away. Toxic levels cause digestive tract and neurological damage, but toxicologists can't say for certain whether lower levels have effects that haven't yet been spotted. Scientists and policymakers therefore lack clarity about where to draw the line between safe and unsafe chronic exposures.[7] Legal limits for

[6] Hicks was officially not a party to the lawsuit, but it rested on BeverageGrades data, and the timing of press releases and marketing materials about the company and the lawsuit made the connection unmistakable.

[7] People who suffer most from chronic arsenic poisoning are also disadvantaged for other reasons. The worst cases in the world happen in Bangladesh and India. Sadly,

safe quantities have dropped multiple times since the alarm was first raised. And yet, even accounting for all of that, wine lovers can rest assured that no matter which wine they pour, they're not pouring too much arsenic into their diets.[8]

The 2015 lawsuit accused eighty-three wine brands, including many that sell in high volumes for low prices, of violating California state law because their products contained more arsenic than the Environmental Protection Agency deems legally acceptable in drinking water. The trouble with that complaint is that wine is not consumed like water—and if you're getting those two beverages mixed up, you have far bigger worries than arsenic. Drinking water in the United States, which is routinely tested for health hazards of many kinds, needs to register under 10 parts of arsenic per billion parts of water. The wines that Hicks's lawsuit named topped out at around 50 parts per billion. In other words, to ingest the same quantity of arsenic from wine as you may legally ingest from your average two or three liters of daily drinking water, you would have to down six bottles of wine each day. That remains true even if we're talking about wine that contains more arsenic than most. A Californian judge recognized this illogic and threw out the case.

The whole debacle was never about arsenic, or wine, or public health. It was about money. Hicks promoted the story to wine writers with an offer to share the data—if they paid for his BeverageGrades service. On the day the lawsuit was filed, wine retailers received emails from BeverageGrades with an offer to verify that the wines they sold were safe—if they paid for the service.[9] Spokespeople for the lawsuit said that they wanted the wine industry to provide consumers with more detailed nutrition information, a valid point. But the arsenic attack itself wasn't, and yet it's persisted in a series of poorly researched news stories that scare people who can't possibly be expected to know better.

---

severe and chronic arsenic poisoning in Bangladesh has followed a shift to obtain drinking water from shallow tube wells instead of surface water. Surface water is often contaminated with disease-causing microbes, so wells seemed like a positive development. Unfortunately, water from tube wells is often disastrously high in arsenic, and improving one life-threatening problem has caused another.

[8] The United States doesn't regulate arsenic in wine or other foods, at least in part because Americans consume specific food products in such radically different quantities.

[9] Frank, "Mixed Case."

In the years since all of that happened, "clean" brands have emerged, warning health-conscious drinkers about the potential dangers of ordinary wines in the interest of selling their own alternatives. As in the arsenic case, many of those so-called dangers aren't a problem in "ordinary" wines in the first place. Added sugar? Illegal in many regions and used in others *before* fermentation to boost alcohol levels in chilly vintages. Any sugar left in table wines is the small, usually nutritionally insignificant fraction that yeast didn't ferment. 🍇 Artificial color? Illegal; while the color of commercial wines is sometimes bolstered with a grape-derived concentrate called Megapurple, or with another densely colored wine, all colors in wine derive from grapes. Low-calorie? A wine's calorie count is a direct function of the alcohol and residual sugar it contains; dry, lower-alcohol wines will be comparatively diet-friendly in contrast to big, ripe clunkers. "Lower-calorie" is just another way to market wines of modest alcoholic strength. The only way to further reduce calories is to artificially remove some of their alcohol, which involves more processing, not less. ⚘ Artificial additives? The list of what can be added to wine is tightly regulated and short compared to what's legal in food: grapes and parts thereof, oak and parts thereof, 🛢 yeast nutrients and acids that are already present in grapes or wine, sulfur dioxide, a few other less frequently used preservatives (far fewer than those permitted in food), a few enzymes added in small quantities in service of wine clarity or microbial stability, and the fining agents we've already addressed that don't remain in the finished wine.[10] Natural wine proponents lambast these as injurious to a wine's soul—an entirely different kind of argument—but, allergies aside, they won't hurt a wine lover's body (see the NATURAL WINE box). Artificial sweeteners, natural and artificial flavors (other than those from oak, which are effectively grandfathered in), and myriad additives that worry consumers in food are never allowed in wine. And for reasons that we've already discussed, clean brands can't assure you that they'll be less likely to provoke a headache than others.

---

[10] Enzymes are proteins that living things employ to make biochemical reactions happen far faster than they otherwise would. In general, the ones used in wine are ordinarily produced by microbes, which use them either to access additional food sources or as defenses against other microbes.

NATURAL WINE

In another instance of confusing terminology, natural wine and clean wine have nothing whatsoever to do with each other. Clean wine is about (mostly specious) health claims, primarily related to what is absent from the finished product. Natural wine is about a commitment to particular production tenets, largely related to how the product is made. Clean wine brands rarely emphasize their provenance, and some don't specify origins beyond broad statements such as "product of France." Some clean wine brands are marketed by celebrities, and some are pyramid schemes that invite you to have parties to sell clean wine to your friends. Natural wines are typically produced by small operations for which connecting consumers to the stories of individual people and vineyards is part of the point.

Clean wine and natural wine are also both distinct from minimal-intervention winemaking, an underspecified term that sometimes (not always) has more to do with marketing than specific commitments in the winery. 🍇 And none of these fuzzy designations has anything whatsoever to do with the various sustainability badges that bottles sometimes bear, which do generally have specific meanings but which don't generally relate in any way to quality characteristics of the wine. It's no wonder at all that confusion abounds.

After many years without one, "natural wine" does, as of April 2020, have a specific definition, though only in France: wine made with hand-picked, certified organic grapes, without the addition of commercial yeasts or otherwise permitted concentrated colors, acids, tannins, sugar, or water. Sulfites mustn't cross the very low threshold of 30 mg/L at bottling, and wines with no sulfites at all have their own additional designation. As one might expect, physical manipulations such as reverse osmosis are also prohibited. Those are stringent requirements (though difficult to verify), but they still might not satisfy those who think of natural wine as a philosophy rather than a prescription.

Nevertheless, some names on the list of what *is* permitted in wine can look scary, so let's consider a few in more detail. Clean wine companies are all too happy to tell consumers that the FDA permits wines to carry as much as 350 parts of sulfites per million parts of wine, and that that's a lot, and that theirs contain far less. What that pitch doesn't mention is that virtually everyone's wine contains far less than the legal limit, because that limit doesn't reflect contemporary winemaking practices. 😒 Clean marketing also tries to induce worries about additives with disconcerting chemical names like tartaric acid, without mentioning that while winemakers can indeed use it to achieve a more microbially stable pH, grapes themselves are the world's primary source of tartaric acid. Similarly, while lysozyme might sound like something you don't want to eat, this enzyme—which winemakers can use to stave off unwanted malolactic fermentation 🍷—is found in your own tears and saliva, as well as in egg whites, where it does essentially the same job for which winemakers employ it: blocking bacterial activity by damaging bacterial cell walls.[11]

The confusion these marketing strategies leverage is understandable. For folks who haven't studied food chemistry, "tartaric acid" and "lysozyme" may sound less natural and more potentially hazardous than "natural flavorings" or "added vitamins and minerals," which winemakers *aren't* allowed to add.

Contrary to what current messaging sometimes suggests, more additives were present in wine before the twentieth century. Lead, for example, was a common (if often clandestine) way to sweeten sour wine from Roman times into the late 1800s.[12] Recipes for wine over the same span might call for adding lead, herbs, salt, honey, or powdered lime to the basic grape-derived product to prepare it for consumption (see Figure 15.1). (That says something about changing tastes and norms, and no doubt about the quality of everyday plonk too.) Lead remained a real live concern for European wine consumers into the nineteenth century, even though a German physician named Eberhard Gockel made the

---

[11] Lysozyme is only active against gram-positive bacteria, which have structurally distinct cell walls compared to gram-negative bacteria, because it degrades cell wall components found exclusively in the former. Conveniently, the cells of mammals, yeast, and other eukaryotes don't have cell walls at all, so lysozyme doesn't affect them.

[12] Lead tastes sweet. Please take my word for it.

Figure 15.1 Wine was mixed with myriad non-grape substances, including lead, until eighteenth-century analytical advances enabled conclusively catching cases of adulteration. Fifteenth-century engraving of "liming the wine," reproduced in Saltarelli, "Les vins des papes d'Avignon et la *colica pictonum* du Vicomte de Turenne." Image in the public domain.

connection in 1696 between wine-related lead consumption and *colica pictonum*—an incredibly painful condition that begins with fatigue and restlessness, proceeds to severe constipation and nervous system damage, and ends in death.[13] Early legislation against "leading" wines (and against adulteration generally, which was widely prohibited long before 1696) was difficult to enforce until analytical chemistry techniques improved in the eighteenth century, after the practice died out.

Excess?

Wine's substantial long-term health concerns, and most of its short-term ones for that matter, are a function of what's *supposed* to be in

---

[13] Eisinger, "Lead and Wine."

wine, not contaminants. Alcohol is behind something on the order of 261 deaths per day or 95,000 deaths per year in the United States alone that wouldn't have happened otherwise, according to the US Centers for Disease Control and Prevention. The World Health Organization estimates that 4 percent of all deaths, globally, are directly attributable to drinking. Most involve consequences of chronic drinking, including liver disease and malnutrition. Yet nearly half of alcohol-related deaths are acute situations: drug overdoses in which alcohol was a factor, car accidents, and suicides, among others. Sadly, those acute issues disproportionately affect young people.

Worldwide alcohol abuse clearly can't be pinned on wine alone. On the contrary, the citizens of traditional wine-consuming countries have been steadily consuming less of it, which sounds like a bad thing for the industry if you don't realize the kinds of per-person volumes we're talking about. As a nation, France is responsible for the second-highest wine consumption in the world, at about 24.7 million hectoliters in 2020, following the United States at 33 million hectoliters in the same year.[14] France, however, is home to not quite 67.5 million people, while the United States has a population almost five times larger at just about 330 million. In the 1950s, the French government attempted to advocate that citizens drink less than a liter of wine per day. In 2019, a campaign from Santé Publique France advertising that two glasses each day was enough was widely criticized as too restrictive, even though that recommendation followed a report citing alcohol as the second-leading cause of French mortality just behind smoking, responsible for 110 deaths each day.[15]

Talking about drinking too much begs the question: how much is too much? Whether or not you're French, I trust that it goes without saying that this question has no one right answer. What doesn't go without saying, maybe, is that scientific research probably can't answer it. What we have right now are coarse guidelines: no more than two drinks or 30 grams of alcohol a day for men, or one drink or 15 grams for women.[16]

---

[14] OIV 2020 annual report.

[15] Santé Publique France, "L'alcool pour comprendre."

[16] These recommendations obviously embed generalizations about men and women that won't apply to many people, but formal alcohol consumption guidelines have yet to respond to changing norms about gender categories and gendered assumptions.

Medical recommendations conclusively agree that no one should begin drinking for their health. The reason they need to bother saying so is that conservative drinking—no more than one drink a day—is associated with a statistically lower risk of dying, from anything, in comparison to total abstinence.[17] These are generalizations about populations. Personalized medicine may, eventually, deliver individualized recommendations about how much alcohol is safe or healthy for *you*. But no matter how good those algorithms become, personalized medical advice will never be able to tell you what you should do without accounting for what you're trying to achieve. And since the data that underpin recommendations are colored by what *researchers* are trying to achieve, we can't ignore that alcohol and health research is always colored by values from the start.

Among the major venues for peer-reviewed scientific research on alcohol and health is the journal *Alcohol and Alcoholism*, published under the aegis of the Medical Council on Alcoholism, founded by members of the British Medical Association in 1969 to "concern itself with the problems of alcoholism."[18] Research in this journal tends to begin from the starting position that alcohol is a *problem* and that the ultimate goal of alcohol-related research is to get people to drink less of it. The *Drug and Alcohol Review* and the *Journal of Studies on Alcohol and Drugs* come from similar angles. The latter was founded in 1940 by Dr. Howard W. Haggard, then director of Yale's Laboratory of Applied Physiology, to support research on alcohol abuse following the repeal of Prohibition in the United States. In contrast, alcohol-related research also appears in such publications as the *Journal of Food Science* and *Appetite*, wherein you'll see wine and beer approached as foods about which consumers need to make nutritionally sound choices. And in journals such as the *European Journal of Preventive Cardiology* and *Pharmacological Research*, wine is addressed as a source of pharmaceutically relevant compounds with potential benefits and harms for people at risk of heart disease, or dementia, or diabetes.

These differences matter to the shape of how science happens, from which studies are funded and what questions are asked to how data

---

[17] Di Castelnuovo et al., "Alcohol Intake and Total Mortality."
[18] "The Medical Council on Alcohol."

are interpreted. A study about European and American consumers' attitudes toward more thorough wine nutrition labeling, for example, begins: "Alcohol misuse ranks among the top five risk factors for disease, disability, and death throughout the world and also has serious social and economic consequences for individuals other than the drinker and for society at large."[19] Research into how drinkers in the United Kingdom respond to labels with lower-alcohol descriptors begins similarly: "Drink-related harm costs the UK government £21 billion a year."[20]

Misuse and harms are one way to frame why alcohol research matters, but not the only way. Studies that frame alcoholic beverages as anomalies in the food world, because they've been exempted from nutrition labeling, begin differently: "Labeling on packaged goods ensures that consumers can make informed purchasing decisions"[21] or "Consumers increasingly demand more information about how their food and beverages are produced."[22] Meanwhile, the first sentence of a report about whether wine drinkers are healthier because they drink wine or because they tend to eat healthy foods begins with a different frame for why wine research matters: "Epidemiological studies have established that moderate alcohol consumption is associated with reduced risk of cardiovascular disease (CVD) as well as multiple health conditions."[23]

Even the most conscientious scientist colors their research with what they think about alcohol, whether their position comes from their mother, their church, their personal history, or their government (see the SCIENTISTS box). And it matters. Research about health hazards often lumps all alcohol consumption into one category; the focus isn't on what or how someone drank, but whether their drinking is high, moderate, or low. (Exactly what those categories mean in any given research context is its own very slippery kettle of fish.) Some but not all ask drinkers separately about wine, beer, and spirits. Cardiovascular researchers may ask separately about red wine versus white (because

---

[19] Annunziata et al., "Do Consumers Want More."
[20] Vasiljevic et al., "Impact of Low Alcohol."
[21] Pabst et al., "Consumers' Reactions."
[22] Streletskaya et al., "Absence Labels."
[23] Hansel et al., "Relationships Between Consumption."

SCIENTISTS

Scientists don't directly make some of the most important research decisions because research requires money. Funding is competitive and increasingly hard to come by. To win it, researchers' plans need to clearly match the priorities of the organization that holds the money. The *Journal of Studies on Alcohol and Drugs* has sometimes been supported by the National Institute of Alcohol Abuse and Alcoholism, a section of the US National Institutes of Health. Recent research into potential health benefits of a wine-inclusive Mediterranean diet has been funded by the health ministries of Mediterranean countries, sometimes with support from olive oil and nut industries, whose products are a prominent part of that diet.

Researchers submit, to one of these funding agencies, a (typically extensive) proposal detailing what they plan to do and what value they expect to deliver as a result. Then they wait for months to find out whether they've made a sufficiently compelling case. While they wait, the agency decides how well the proposal fits within the scope of what they're willing—or able, especially in the case of public funders constrained by larger agendas—to support. The percentage of proposals eventually supported by any given scheme can be in the single digits. Consequently, mountains of well-considered, potentially valuable science doesn't happen because the topic or approach isn't the highest priority in light of limited resources. The result is that the science that *does* happens is intensively shaped by the values of the institutions that fund it. But when scientific knowledge becomes baked into public health recommendations as "what the science says," it's easy to lose track of the values that shaped them in the first place.

red wines are generally higher in cardioprotective phenolics; people who study alcohol abuse rarely do. Many abuse-oriented studies differentiate between drinking one drink at a time versus bingeing on several in one sitting.

Research can't speak to phenomena that researchers aren't looking for. Most studies have prioritized the *volume* of alcohol that people consume. Only recently have studies investigating the evident health benefits of Mediterranean and other traditional wine-drinking diets begun to consider that exactly *how much* you drink may be less important than *how* you drink. These investigations suggest that some benefits of traditional food-and-drink habits may stem from eating and drinking in the convivial company of family and friends, not solely the nutritional composition of what's on the table. Heretofore, very few studies have accounted for the possibility that two or three glasses of wine with a long family meal may have different health consequences than consuming the same number of glasses by yourself in front of the TV while having a leftover bagel for dinner. If a study doesn't include a metric to document the difference between those two scenarios, researchers can't identify how their health effects might differ. Those differences, if they exist, become invisible.

Twentieth-century biomedical research (aided by contemporary productivity expectations) has encouraged adults the world over to stop seeing drunkenness as an ordinary part of the working day, whether it's the product of a three-martini corporate lunch or a carafe of wine in the factory canteen. This is obviously a good thing. Less obviously, taxation and other measures aimed at getting people to drink less affect some people more than others. *Mad Men*, a wildly popular and unbelievably soggy TV series about men (and a few women) who drink their way through advertising careers in the 1950s, spawned cocktail books and drink pairings. Move the same consumption habits out of a classy wood-paneled bar into a trailer park, switch the slick suits for sweatpants, and they become a stomach-turning societal disgrace instead of an endearing amusement. Workers having a glass of wine with lunch is a crime against efficient production. Businessmen keeping a bottle of whisky in their desk drawer is kind of cool.

The saddest thing about scientific research on alcohol and health, toxins, sulfur, and even excess may be that it too often seems to lead people to be anxious that they're doing it wrong. I've avoided the common phrase "alcohol use" in this chapter because "use" subtly sends the message that we're talking about drugs. We talk about heroin use and nicotine use. We don't talk about mozzarella use or mushroom use, unless we're making a tongue-in-cheek critique of a friend's pizza habit or complaining about their recipe for lasagna. Anxieties about consumption make us vulnerable to consumerist, profit-driven ploys that offer new products to assuage those anxieties. When we talk about alcohol use, we fall prey to the bizarre idea that wine is a drug. When we become obsessed with what companies aren't telling us, we become susceptible to even more companies that try to claim that they're different.

I recall a comment from a colleague on Twitter who complained, after attending a party the night before, that wine is a terrible drug. Two or three bottles had left him feeling worse than smoking a few joints or snorting a line. Would he have even conceived of the possibility of this kind of excess if he thought about wine as food? No one imagines that eating two or three cartons of ice cream, two or three loaves of bread, or even two or three heads of cabbage all in one sitting could possibly be a reasonable idea. He was doing it wrong.

16

Glass ▮

Most reasons to talk about glass, as a wine container at least, boil down to two molecules: carbon and oxygen. Glass bottles are the single largest contributor to a wine's carbon footprint, and far heavier than alternative containers. They're also wine's best protection from oxygen during storage. But just how much protection does wine need? Is that even the relevant question? The twenty-first-century wine packaging challenge isn't so much about finding a good technical response to a problem as figuring out what the problem is in the first place.

The Great Closure Wars

As the noughties wound down, the wine industry was embroiled in conversations around closures: corks, screw caps, cork-shaped objects made of plastic, and a handful of other variations on the theme of keeping wine inside its bottle until someone wants to drink it. The mainstream Australian and New Zealand markets had already settled on screw caps for all but age-worthy reds and the occasional outlier. The rest of the wine world remained unsettled about the technical adequacy of cork alternatives, and about the stigma they might carry.[1] Endless marketing studies wavered between finding that consumers would only accept screw caps on inexpensive bottles and finding that no one really cared, depending on who was asked and how.

---

[1] The vast majority of New Zealand's exports are fresh and fruity sauvignon blancs, and at that time especially, Australia was known for shipping cheap and cheerful "critter" wines primarily distinguished by the kangaroos and leaping lizards on their labels. The general apprehension was that screw caps might suffice for these wines, but not for more serious offerings.

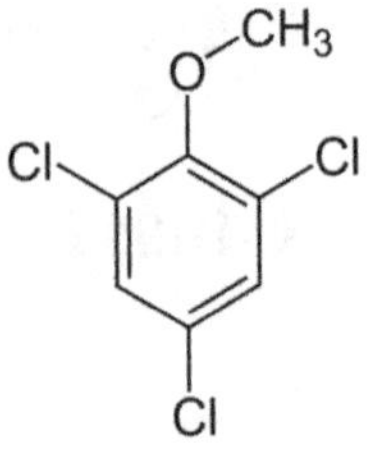

2,4,6-trichloroanisole

Figure 16.1 2,4,6-tricholoroanisole, responsible for cork taint.

Natural cork had tradition and ageability on its side. Like other tree products, 🛢 cork has the rather magical capacity to be solid enough to block the flow of liquids but porous enough to allow the flow of gases. The oxygen ingress that a cork permits is tiny, but key to how bottled wines age.[2] The fundamental trouble with this ancient and wonderful technology is that natural corks also vary in two occasionally ruinous ways. First, individual corks have historically been inconsistent in how much oxygen they let in, such that individual bottles age differently and will occasionally oxidize prematurely. Second, corks are the most common vehicle for cork taint, which also affects individual corks inconsistently. And variability itself is the last thing anyone wants in packaging.

Cork taint is the cardboardy or musty aroma associated with a single powerful molecule, 2,4,6-trichloroanisole (TCA) (see Figure 16.1). It's also associated with the practice of tasting before a bottle is poured at restaurants, since cork taint is the most likely reason why it might need to be sent back. TCA isn't cork's fault as such; it can be found in wines bottled without them, and in products for which cork isn't relevant at all, including water, potatoes, grains, sake, and chicken, among others.[3] The real problem, and the reason why TCA has been such a plague upon the winemaking earth, is that no single factor can be blamed.

---

[2] Scientists disagree about how oxygen enters a bottle, much as scientists disagree over the cognate question for oak barrels. The spongy cork itself contains some oxygen that it will release into the wine. Thereafter, the relative importance of oxygen penetrating the cork or sliding around the cork where it interfaces with the bottleneck occasions some debate.

[3] Hjelmeland et al., "High Throughput."

TCA is one member of a family of chloroanisoles. Chloroanisoles result from bacteria and fungi metabolizing chlorophenols in their environments. Chlorophenols come from lots of places. They were used extensively as insecticides, fungicides, herbicides, and wood preservatives from their introduction in 1936 (thank chemical giants Dow and Monsanto for that) into the 1980s, when they were widely banned for being highly toxic and slow to degrade. Unfortunately, they're remarkably stable in water and soil, so chlorophenols are still routinely found in the bark of cork oaks and in other agricultural products.

Chlorine and phenol can also react to form chlorophenols.[4] Oak products already contain abundant phenols. 🛢 Before TCA's origin story was traced in the late 1980s, corks were routinely bleached with calcium hypochlorite, a source of chlorine, for aesthetic purposes. Both bleach and chlorinated municipal water are now forbidden in cork manufacturing, and cork producers have instituted more stringent antimicrobial protections as well as rigorous checks for TCA and related compounds. Environmental chlorophenol pollution itself will eventually age its way out of tree bark, though it hasn't yet.

Estimating cork taint's prevalence is fraught—since it's not uniformly distributed, testing a random sample of bottles isn't reliable—but its incidence seems to be less than 5 percent of cork-sealed bottles and declining. Unfortunately, once TCA is in a cork and the cork is in a bottle, contamination of the wine is inevitable. Humans can smell as little as five parts of TCA in a trillion parts of wine, and even sub-sensory-threshold amounts can mask a wine's aroma.

Even though they couldn't guarantee a TCA-free wine (since the TCA could come from other sources), screw caps offered more consistency than corks by solidly blocking exposure to outside air. They therefore also blocked the expected aging trajectory of wines built for cellaring. In addition, their tight seal seemed to foster the buildup of volatile sulfur compounds with charming aromas ranging from rotten egg through a range of cooked vegetables, producing a character known as "reduction." 🛢

---

[4] Simpson and Sefton, "Origin and Fate."

Screw caps also came with a risk that had nothing to do with wine quality directly. Marketing research indicated that at least some consumers either consciously saw the closure as lowbrow or unconsciously responded to it as less premium. And making the situation more complex, natural cork and screw cap companies tussled over which was more environmentally friendly, with the available evidence supporting no clear winner on either side.[5] Whether or not this *new* technology was also an *improved* technology was genuinely up in the air for many producers at anything but the bottom of the market.

The middle ground between natural cork and totally uncork-like screw caps was occupied by a variety of "technical" or "agglomerated" corks, molded and extruded from some combination of cork offcuts and binding agents, and "synthetic" corks made from plastic. At their best, these options melded consistency and tradition, sometimes with carefully standardized oxygen transmission rates. At their worst, plastic corks were too inelastic to provide a reliable seal and stupidly hard to extract from the bottle, making cheapness their singular virtue.

Faced with serious competition, natural cork producers began instituting better screening processes for cork taint and uniformity. Screw cap producers engineered slightly porous versions of their products by drilling microscopic holes in the aluminum and selecting plastic liners through which measured bits of oxygen will pass. Meanwhile, winemakers have refined techniques for oxygen-occlusive closures to avoid stinky buildups. Manufacturers of technical closures, trading on the consistency and control they can offer, have worked to educate winemakers on the virtues of managing "total package oxygen." And plastic corks are still cheap. When the American industry magazine *Wine Business Monthly* conducted their annual closures survey in 2021, 70 percent of responding wineries reported using natural corks for at least some wines, but 52 percent use at least some screw caps, and about a third use technical corks. What those numbers don't reflect is

---

[5] Harvesting cork doesn't kill the cork tree, a species of oak called *Quercus suber L.*; on the contrary, cork forests are a major biodiversity haven and the trees have legal protections. Screw caps rely on aluminum. Producing aluminum from bauxite—a mixture of aluminum hydroxide and impurities obtained by surface mining—leaves behind toxic sludge called red mud that, when not properly contained, has caused atrocious and fatal accidents. Manufacturing corks, however, typically requires more water.

just how common Anything But Cork has become for everyday, inexpensive wines, since so many enormous producers have made the switch even while corks remain favored among the boutique crowd.

So, if the closure wars haven't been entirely settled, then at least the conflict has receded enough that all parties coexist more or less amicably. For less expensive wines, closures aren't a focal point. Screw caps are topping enough higher-end wines for any stigma to begin to fracture. At a recent neighborhood wine party, I needed to run back to my place for a corkscrew, because my contribution was the only bottle that required one. (Don't take that to mean that I'm precious about corks. I'm not, but many natural wine producers prefer natural corks, and I prefer many natural wines.)

Today, we're seeing a different fight. The closure wars assumed that the bottle itself was a given. These days it's not. Every bottle on the table at my community's party was indeed a bottle, but the local shop also sells canned wine, boxed wine, and Tetrapacked wine. (Tetra Pak is the dominant brand name behind the laminated paper cartons in which you might buy oat milk or single servings of juice.) A larger and winier town would also offer wine on tap to be carried home in a growler, and maybe some options in plastic or paperboard bottles. Alternative containers seem poised to follow the same trend over the next ten years that alternative closures have followed over the past ten, driven by similar technical developments and the ultimate economic and environmental unsustainability of glass.

The Problem with Glass

The predominant reason why anyone wants to give up glass is that it's heavy. Excepting a few mega-wineries that manufacture bottles onsite, producers must ship them twice: empty, then full. Shipping heavy things is expensive and carbon-intensive. Between transport and the energy-demanding process of making the glass in the first place, bottles account for the largest proportion of a wine's carbon footprint, according to numerous studies across multiple countries that account, in some cases, for details right down to the oil that greases the wheels of the tractor that pulls the cultivator that weeds the vineyard. For heavier bottles, transport alone is the lion's share. A freight truck carrying wine packaged in thick bottles will exceed its maximum weight capacity before it fills up on volume.[6]

Lighter bottles would seem like a no-brainer. Regrettably it's not, because wine producers and wine consumers have been caught in a circular marketing argument that equates weight with quality. Research from the Oxford-based Crossmodal Sensory Lab ♪ says that consumers will pay more and think that they receive more value for wine packaged in heavier bottles—a classic example of sensation transference, or the tendency for one characteristic of an object to influence what we think about unrelated characteristics.[7] Responding to that tendency, the industry has bestowed on consumers opportunities to purchase bottles weighing nearly two pounds, with the rare prestige example even exceeding the two-pound mark. A bottle half that weight is equally good at containing the wine inside.

This situation is dreadful. Consumers pay extra for unnecessary packaging—£1 for every 8 additional grams of glass, in an analysis of wines sold in Oxford.[8] Toting that packaging around generates senseless pollution. And yet if anything, bottles have become heavier over the past several decades, with producers trapped in an arms race of trying to stand out amid an ever-growing sea of similar bottles, and bottle weight an expected part of the arsenal for doing so. Bizarrely, it's not clear why. Research clearly shows that bottle weight and wine cost are correlated, but we have no data to say whether consumers will reject expensive wine in light bottles. Marketing researchers can't just go out and ask them, because what people say they care about under experimental circumstances is quite different from what they and their wallets do in real life. Understandably, few wine producers have

---

[6] The combined space and weight constraints of a semi-truck mean that producers benefit much more in switching from a heavy bottle to a midweight one than from a midweight bottle to a light one. A truck carrying full midweight bottles will reach its weight maximum at about the same time as it runs out of volume. With light bottles, when the truck is maxed out on volume, it will still have weight capacity to spare.

[7] Kampfer et al., "Touch-Flavor Transference." The same principle applies to perceiving the quality of other foods; in one experiment, for example, tasters liked yogurt more when it was served in a heavier bowl than a lighter one.

[8] Piqueras-Fiszman and Spence, "The Weight of the Bottle as a Possible Extrinsic Cue."

jumped at evaluating whether consumers will (consciously or unconsciously) discount lighter bottlings in practice.

As tempting as it might be to say that humans are hard-wired to prefer heavier products, we don't equate heft with quality in all cases; for plenty of luxury goods, including high-end laptops and sunglasses, lighter is better. Producers and consumers alike are caught in a positive feedback loop of learned responses. More expensive wines come in heavier bottles, so drinkers learn to expect heavy bottles for expensive wines, and producers learn that lighter bottles make wine seem cheap. And so here we are in the 2020s, with glass manufacturers offering new designs on glass doorstops that also happen to function as wine containers.

That's the bad news. The first bit of good news is that purchasing data indicate that younger, more eco-minded drinkers are already learning to prefer lighter bottles. The second bit is that some wineries are beginning to switch to lighter bottles simply because they're cheaper. We can only hope that those changes will provide openings for further change, and for observing that maybe consumers don't care as much about packaging as producers have feared—much as happened with screw caps on more expensive wines.

Eventually, there comes a point at which lighter bottles become easier to break. The glass industry has been developing technologies for maintaining strength while "light-weighting," including blowing and molding refinements that can make a bottle's thickness more uniform (a factor in breakability), redistributing glass from the bottleneck to other locations where it does more good, and coatings that reduce breakage. In other food and drink industries, technical developments have lightened packaging by at least 50 percent compared to mid-twentieth-century weights. Those developments have also been relevant for the cheapest wines. But the wine industry's challenge is primarily social, not technical. Wineries need to be convinced that light-weighting won't affect their image, and consumers need to be convinced not to be so affected.

Is Glass the Problem?

But maybe we've been addressing the wrong problem. Maybe it's not that bottles are too heavy, but that they're bottles. Maybe glass isn't the best tool for the job and alternative containers are the answer. The incongruous thing, given the tenor of most sustainability conversations, is that alternative containers come down to plastic, even when they don't seem to.

Take bag-in-box wines. The plastic bag is all that matters to this old stalwart of the not-a-bottle brigade. When first developed in 1965, bag-in-box wines were just bagged wines.[9] The box was added because liquid-filled plastic bags are about as easy to handle as a live cephalopod, and because boxes' large, flat surfaces are a boon for marketing content. The cardboard does absolutely nothing from a wine-quality perspective; the plastic alone constrains how well the wine keeps.

Until recently, your standard wine bag was always assembled from several layers of plastics. Some were polyethylene, the world's go-to plastic from which grocery bags, water bottles, and much of the other packaging in your house are formed. Annoyingly, polyethylene breathes—not so well that you can breathe through it, mind, but enough so that you'll see plastic-wrapped foods begin to oxidize and turn brown within a few days. Consequently, to block oxygen, bags need a middle layer of metalized polyester—better known as Mylar—that doesn't breathe, but that also doesn't touch the wine, because corrosion-prone metal in direct contact with acidic wine is bad.

The trouble with laminated plastics is that they delaminate when handled. When layers separate, their structural integrity is damaged, and ambient air with its 20.95 percent oxygen is bound to get in. The plastics industry is solving this problem. Newer polymers can block oxygen more resiliently with just a single layer. No matter how good the bag becomes, though, air will enter around the spigot that lets wine out, giving bagged wines a life span of few weeks at best after they're opened—far better than a bottle with its macroscopic hole in the top, but they won't last forever. Unopened, even improved plastic bags will let in measurable oxygen within about three to eight months, depending on how well the bag-spigot interface has fared during the

---

[9] The story goes that Tom Angove and the first companies to improve on his initial design repurposed bags designed to contain battery acid—welcome reassurance that the acidic wine wasn't going to eat its way through the bag.

bag's travels and the temperatures it's seen along the way.[10] Trained tasters notice declines in bagged wines stored for only a few months under less-than-ideal conditions. Whether you'll notice the same is, of course, a different question. ک

Meanwhile, canned wines have been improving for the very same reason as bagged wines: plastics. The first canned wines appeared in the United States in 1936, very soon after canned beer. These initial efforts were plagued with spoilage issues: balls of fungus growing inside, hazy microbial spoilage, and corrosion severe enough for cans to leak.[11] While the canning process had obviously yet to be refined, part of the problem was the wine. At the time, most Californian wine was sweet and prone to microbial spoilage no matter how it was packaged. Worse, highly acidic wine (because adding acid mitigated spoilage) with lots of oxygen (because oxygen management wasn't what it is today) may as well have been custom-designed to eat through cans protected by the rudimentary protective liners available in the 1930s (see the ALUMINUM box). The wine industry took decades to properly recover from that experience, but it's no wonder that cans have made a comeback: they're portable, convenient, easy to carry up a

---

ALUMINUM

Aluminum is highly reactive. That's rarely obvious precisely because aluminum reacts so fast. Upon exposure to air or water, a nanoscale surface layer of the metal oxidizes to aluminum oxide (there goes chemistry jargon being logical again). That surface, now highly stable, protects the rest of the metal from reacting, such that a ball of aluminum foil will sit unchanged on your kitchen counter just about indefinitely. However, the aluminum oxide layer will happily react with acids to form powdery derivatives that expose more and more of the metal underneath, eating their way through and totally defeating the purpose of using aluminum as a container, which is why bringing it into direct contact with wine is a bad idea.

---

[10] Doyon et al., "Canadian Bag-in-Box."
[11] Berti, "Effect on Wine."

mountain in a backpack, not prone to leaving stabby shards lying around festival grounds, and photogenic to boot. (Caveat emptor: wine marketing research says that canned wines sell best to millennials and Gen Z-ers, and I'm a millennial. I think they're great.)

Even though they appear to be entirely aluminum, cans must be lined with a micrometer-thin plastic film to avoid unpleasant reactions. (The lining is melted off when the can is recycled.) As researchers from Cornell's oenology department have pointed out, canned wines aren't contained in cans at all, but in ultra-thin plastic bottles wrapped in aluminum coatings for the sake of structural support and improved impermeability—bags in cans, rather than bags in boxes.[12] Unlike cardboard, that metal support is also essential to preserving the wine. Thin plastic is rubbish at blocking oxygen. Properly lined aluminum, as screw caps show, blocks it completely.

So once again, just as for screw caps, reduction is the bigger concern with cans. In the presence of acids, sulfur dioxide will react directly with aluminum to form rotten-egg-redolent hydrogen sulfide and water. 💿 Plastic liners *should* prevent this problem. The trouble is that scientists have seen sulfur dioxide diffuse through plastic films. Microscopic imperfections in liners will allow for direct contact, too, and are inevitable to some degree. Keeping sulfur dioxide to a minimum in canned wines helps, and is less risky than it might be in a cork-sealed bottle precisely because canned wines are less threatened by oxidation, but the risk can't be avoided entirely.

Oxidation also isn't a total non-issue. Cans, like bottles, need to be sparged with an inert gas such as nitrogen to force air out of the container before it's filled, lest atmospheric oxygen damage wine when the two are trapped together in close quarters. (That's one reason why cans are brilliant for sparkling wine, because sparkling wine sparges itself.)[13] A poorly calibrated canning line may sparge inconsistently. Air can also seep through the join between can bodies and can lids, though that possibility is insignificant in light of current can designs so long as

---

[12] Allison et al., "The Chemistry of Canned Wines."
[13] Canned sparkling also pays off the most in sustainability terms, since bubbles offer the one good excuse for a heavier glass bottle; with dissolved CO_2 pushing outward at six times atmospheric pressure, they legitimately need the extra strength.

canning equipment is performing correctly. Wine can manufacturers warranty their products for a year. Canned wines may keep for far longer, though a great deal rests on the integrity of that liner, reduction is most likely to be the limiting factor.

Problems with Plastic

Unless you are so very old-school as to package the fruits of your own labor in animal-hide wineskins, plastic is involved in every bottle alternative on the market. Bags are plastic. Cans are aluminum lined with plastic. Tetra Pak cartons are paperboard sandwiched between a layer of outer plastic, a layer of inner plastic, a layer of aluminum to block oxygen, and two more thin layers of plastic to protect the aluminum. A few bottom-shelf wines are packaged in plastic bottles. Wines on tap at a bar or restaurant come from kegs that use a scaled-up version of bag-in-box or can technology, all relying on plastic.

Plastic isn't inert. Polycarbonates, the garden-variety plastic common in food and beverage packaging, have typically been produced by cross-linking molecules of a petroleum-based ingredient called bisphenol A (BPA) to each other with the aid of a chlorine-containing gas called phosgene (see Figure 16.2). Phosgene is highly toxic and has caused fatal accidents in the vicinity of plastics factories, but isn't present in the finished product. BPA, on the other hand, has

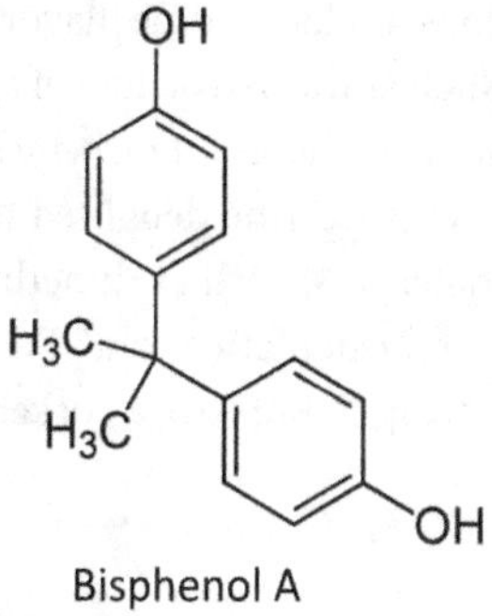

Bisphenol A

Figure 16.2 Bisphenol A.

become a household name because it can seep from finished plastics into foods and drinks and thence into humans.

BPA weakly mimics estrogen. Lab rodents fed improbably large doses of the stuff will develop abnormal reproductive organs and neurological problems. As with most such studies, conducting similar experiments in humans is obviously unethical, and extrapolating from high, acute doses in rodents to lower, chronic doses in humans is full of uncertainty. Nationally sponsored research in the European Union and United States attests that a typical citizen's BPA exposure is well within safe limits. Nevertheless, major wine bag and can manufacturers have replaced BPA with less toxic alternatives, in line with actors in other industries. Still, since wine might absorb some BPA from other plastic contact points during processing, no one can guarantee that their wines are entirely BPA-free.

BPA aside, plastic is an issue not for what it might add, but for what it might remove. Plastic "scalps" flavors, an unpleasant but exceedingly common way to talk about the chemical phenomenon of adsorption. Adsorption is similar to absorption. In both cases we're talking about one thing being taken up by another. Adsorption, however, is the specific term for molecules adhering to a surface. Lint adsorbs to corduroy trousers. Thirsty people adsorb to the outward face of a bar. Highly uncharged molecules will adsorb to a can's plastic liner, or to the inner surface of plastic bags, plastic liners under aluminum screw caps, and plastic cork-like closures, trapping them in the package and keeping them away from your nose and mouth.

Felicitously, wine contains very few highly uncharged molecules. (Uncharged molecules dissolve poorly in water, and therefore also in wine. ⋰⋰.) Exceptions include some flavor compounds that we wouldn't want to lose, such as the petrol note of aged rieslings.[14] On the whole, those are a minor consideration for wines built to drink young. And felicitously again, we might be delighted to lose a few other exceptional uncharged molecules, such as dimethyl disulfide, a garlicky sometimes-byproduct of fermentation, and TCA. 🅢 The old trick of using balled-up plastic wrap to improve a corked wine has some merit for the very same reason.[15]

---

[14]   Allison et al., "The Chemistry of Canned Wines."
[15]   González-Centeno et al., "Use of Alimentary Film."

Plastic is a trade-off. Bagged wine weighs only about a quarter of what an equivalent volume of bottled wine typically does, and so saves energy on shipping. It's also a petroleum product, and rarely recyclable on account of being tied up in composite materials; even single-layer bags have that troublesome plastic spigot. The plastic liner is melted off cans when they're recycled, and consumers recycle aluminum cans more often than either glass or plastic, but cans use more packaging overall because the packages are smaller. Recycling itself is inefficient in that most recyclable waste isn't recycled, and the process requires energy and water.

In the end, packaging trade-offs are a game of efficiencies and shifting local circumstances.   For the time being, the carbon-footprint analysts say that bagged wines are the most efficient choice, but only for wines intended for rapid consumption. Throwing away spoiled wine is efficient for no one.

Good Enough?

From a wine quality perspective, there's no contest: glass wins at being protective. Alternatives win on weight but remain lacking on the protection side. The thing is, considered in social context, glass is a silly gold standard. The biggest concern for wine packaging is whether wineries and their patrons will buy it. Once they do, preserving power rarely makes a difference.

Ninety percent of wine purchased in any kind of takeaway container in the United States is drunk within two weeks of being brought home.[16] For that 90 percent, glass is overkill, far better than it needs to be to do its job. Let's imagine that alternative packaging is for wines meant to be drunk young, just as screw caps are great for most wines without always being the best choice for aging. The benchmark for success becomes whether they'll keep wine fresh for its journey from the winery to the shop and a bit thereafter, not whether alternatives can replace the bottles in aficionados' cellars.

---

[16] Thatch and Camillo, "A Snapshot."

At the end of the day, "wine" is too blunt a discursive instrument for evaluating packaging, which makes packaging like most other wine considerations. Making sensible choices—economically, environmentally, and hedonically—requires distinguishing wines that will be lovingly stored and enjoyed years or decades after bottling from wines whose best-by date might be eight months after fermentation finished. These are two different animals entirely.

That difference, more than anything else, is why we're already setting up to skip the Great Container Wars. Unlike closures, none of the current range of technically adequate, commercially viable container options is being positioned a one-to-one replacement for another. I can't help but hope that the default container a decade from now is the can (for everyday sparklers and small portions) or recyclable plastic (for still wine in larger volumes) or refillable glass (the best option of all for folks who live reasonably close to wineries or in towns where popular wines turn over fast), plus traditional bottles for wines built to age. Give us a few decades of that kind of differentiation, and we may remember the era of the default glass wine bottle as a peculiar anomaly.

Coda

I've called this a coda, not a conclusion, because it makes a point that I hope is obvious by now but that I'd feel irresponsible not stating outright. Boring wine can't be blamed on science and technology. Breathtaking bottles can be made with new methods. Winemakers benefit from understanding the mechanics behind the processes they employ regardless of the styles they produce. The quality of everyday wine, on the whole, has improved. Wine science *has* sometimes been presented as a set of instructions for making standardized, "safe" wines. But that's more a fault of rigid mindsets than of molecular biology.

The trouble with talking about relationships between science and wine is that "wine" is an absurd category. It's good enough as a designation for the alcoholic product of fermenting fruit juice: grape wine, raspberry wine, mango wine, and so on. It's crummy as a meaningful descriptor for how those products relate to people and places and experiences. Some wines are adequate beverages and nothing more. Some wines are sublime expressions of beauty and awe. Plenty are somewhere in between, lively and interesting without compelling the center of attention.

Distinguishing the functional and technically adequate from the expressive or artistic isn't a wine-specific issue. Renowned director Martin Scorsese recently made a similar distinction about film to defend his polarizing comment that the Marvel Cinematic Universe isn't cinema. Cinema, he argued, is an art form that takes risks in the course of saying something about human experience. Film products in Marvel's line, in contrast, deliver formulaic entertainments. They can be perfectly good entertainment, but they're not art. Cinema may not always succeed at being art, but at least it has a shot. We could have the same discussion about books (literary and commercial fiction), clothing (couture and body coverings), home furnishings, restaurants, and so on.

The same scientific finding can be taken up and applied as a formula to produce a formulaic wine, or as an enabler of other goals. The difference isn't about good and bad or better and worse. Quaffable entertainment is sometimes just the thing. Couture isn't built for everyday wear. I sometimes want a glass of solid table wine that doesn't ask for much energy at the end of the day, when something grander would be out of place.

Science and technology span the spectrum rather than coming down on one end. Minimal-intervention winemaking can yield engrossing wines, quaffable wines, boring wines, or faulty ones. Engrossing wines can be produced using technologies developed well before the twentieth century or technologies developed after it. Wine science has reshaped distinctions between *vin ordinaire* and fine wine, but certainly didn't invent them.

And yet there's obviously a relationship here, which I think is why this whole non-equation of art and science becomes so muddled. Big-budget films and big-batch wines need to ensure that the big investments they require pay off in robust sales. Their producers mightn't be able to afford risks available to indie folks with less to lose or more modest expectations, maybe driven by a personal vision more than a corporate agenda. Beyond making a product with mass appeal, big-brand winemakers may need to intensively engineer that product for consistency from batch to batch. Small producers often (though certainly not always) can't afford the same kind of tech that the mass-market folks can, and won't benefit from the same economies of scale, but are more likely to have customers who expect year-to-year variation. Scale matters to what makes sense, and cents. So do imagination and risk aversion. Economic and marketing relationships are at the heart of who's most inclined to use science formulaically, in addition to personal predilection. So of course it's easy to misidentify, and to think that technoscientific interventions make boring wine (or films), even when a more detailed look clarifies that that's not true.

The muddled relationship between science and boredom brings us back to where we started: values. Science doesn't magically avoid entangling in values. Researchers research for reasons, with funding attached to public or private agendas; individual and institutional perspectives, assumptions, and beliefs become baked into their findings. It's entirely possible to value science without agreeing whole-sale with the assumptions and values baked into it. It's entirely possible to learn about wine research and take it into account without embracing goals identical to those that initially motivated the research. And winemakers often do.

It's an oft-repeated truism that we wine-loving inhabitants of the twenty-first century have access to better-quality wines, on the whole, than our great-grandparents did. Records left by wine scientists a few generations back attest that refermentation in the bottle, weird micro-bial spoilage, moldy corks, haziness caused by unappetizingly high mineral and protein concentrations, and other unpleasant or ined-ible situations were comparatively common not long before our cur-rent winemaking era. My older professors had firsthand experience with faults that are now rare on the commercial market. A global shift from producing oceans of wine for consumption with every meal to-ward connoisseurship has opened more space for prioritizing quality over quantity. Emile Peynaud, who died in 2004 after a long career as a Bordeaux-based wine professor, counted his most significant contri-bution as convincing French winemakers to vinify only ripe grapes, excluding the underripe, overripe, and spoiled ones previously in-cluded in the name of economy.

But wine science can't take credit for all of that. No disrespect to Peynaud, but shifts in how wine is marketed and sold—pressures on European producers from escalating New World competition, Americans' love of numerical scores, expansion of the global market, and so on—probably carried more weight than he did.[1] Scientific re-search has been an enabler, providing guidelines for how to avert faults and achieve particular styles, but always in a wider context that grounds the goal of the enterprise.

Several of my favorite winemakers talk about how long they took to recover from what they were taught en route to their viticulture and/or oenology degree. Now, appreciating the lifestyle preferences of lactic acid bacteria or biochemical details of how sulfur dioxide

---

[1] As William Echikson recounts in *Noble Rot*, Peynaud's career coincided with changes, including the rise of Robert Parker, that pressured prestigious estates to im-prove quality.

interacts with oxygen hardly keeps them from crafting wine as they want it to be. No, the problem is that when those ideas are taught, they're often embedded in the assumption that lactic acid bacteria *need* to be inoculated to ensure that malolactic fermentation is properly controlled, and that sulfur dioxide *should* be adjusted to standardized, safe levels.

"Safety first" pays as a research motto. Bottling an expression of human experience sometimes requires risks. But we can't just lay this problem at the feet of safety versus risk-taking. Across most of contemporary Western society, science has taken point position as the most authoritative way to make knowledge about the world, so comprehensively that it's easy not to notice the hierarchy anymore. When a sentence begins with "The science says . . . ," what follows is usually set up as what we *should* be doing. Regrettably, conflating knowledge about how things work with recommendations about what to do about them has now contributed to an absurd and horrific polarization between science-believers and science-deniers. In an innumerable-orders-of-magnitude-smaller problem, it's also made extra work for some institutionally educated winemakers.

Science is one set of ways of knowing, not universally the best or most important one. Science is a process that changes as it goes. And most of all, science comes from locations in geography and history, space and time, that shape it, such that scientific knowledge comes from a context and needs to be interpreted in context, including the context of what you're trying to achieve. Acknowledging context means seeing science as incredibly useful, but not the only valid way of knowing, nor all-powerful in offering solutions. In a world in which it's increasingly difficult to ignore that all people have a necessarily limited perspective, this attitude is becoming more common.

These shifts are wonderful news for wine. We walk into wine shops and find a gazillion styles, rooted in all manner of notions about good wine, including reliably clean, consistent, high-volume brands. Strikingly few suffer from overt microbial spoilage or profound chemical imbalances. Old-fashioned mindsets about science, a focus on fault aversion, economies of scale, mass-market measures of success, and other historical contingencies of how wine and science meet have indeed sometimes colluded to equate a "scientific approach" or "following the science" with making humdrum wines. Those are contingencies, not inevitabilities. And they're changing. We're firmly in the midst of a different era, in which wine science is overtly more about options than prescriptions, (gradually) becoming more inviting toward diverse perspectives and values, and growing generally more multiple all over. If, like me, you use "interesting" as a compliment, these are interesting times.

References

Allison, Rachel, Gavin Sacks, Luna Maslov-Bandic, Austin Montgomery, and Julie Goddard. "The Chemistry of Canned Wines," n.d. Research News from Cornell's Viticulture and Enology Program. Appellation Cornell.

Alston, Julian M., Kate B. Fuller, James T. Lapsley, George Soleas, and Kabir P. Tumber. "Splendide Mendax: False Label Claims About High and Rising Alcohol Content of Wine." *Journal of Wine Economics* 10, no. 3 (2015): 275–313.

Amerine, M. A. "Hilgard and California Viticulture." *Hilgardia* 33, no. 1 (1962): 1–23.

Amerine, M. A., and A. J. Winkler. "Composition and Quality of Musts and Wines of California Grapes." *Hilgardia* 15, no. 6 (1944): 493–673.

Amerine, M. A., and A. J. Winkler. "Composition and Quality of Musts and Wines of California Grapes." *Hilgardia* 15, no. 6 (1944): 493–673.

Annunziata, Azzurra, Eugenio Pomarici, Riccardo Vecchio, and Angela Mariani. "Do Consumers Want More Nutritional and Health Information on Wine Labels? Insights from the EU and USA." *Nutrients* 8, no. 7 (2016): 416.

Antunes, Francisco de Jesus Chaveiro Ribeiro. "Consumer Preference for Warm or Cold Climate Wine Styles Is Dependent on Emotional Responses and Familiarity." Master's thesis, Universidade de Lisboa, 2018.

Atkin, Tom, Damien Wilson, Liz Thach, and Janeen Olsen. "Analyzing the Impact of Conjunctive Labeling as Part of a Regional Wine Branding Strategy." *Wine Economics and Policy* 6, no. 2 (2017): 155–64.

Ball, Vaughn C. "Shaping the Law of Weather Control." *Yale Law Journal* 58, no. 2 (1949): 213–44.

Berti, Leo A. "Effect on Wine of Type of Packaging." *American Journal of Enology and Viticulture* 1, no. 1 (1950): 119–23.

Bioletti, Frederic T. "Outline of Ampelography for the Vinifera Grapes in California." *Hilgardia* 11, no. 6 (1938): 227–93.

Bokulich, Nicholas A., John H. Thorngate, Paul M. Richardson, and David A. Mills. "Microbial Biogeography of Wine Grapes Is Conditioned by Cultivar, Vintage, and Climate." *Proceedings of the National Academy of Sciences* 111, no. 1 (2013): E139–48.

Bokulich, Nicholas A., Thomas S. Collins, Chad Masarweh, Greg Allen, Hildegarde Heymann, Susan E. Ebeler, and David A. Mills. "Associations

Among Wine Grape Microbiome, Metabolome, and Fermentation Behavior Suggest Microbial Contribution to Regional Wine Characteristics." *MBio* 7, no. 3 (2016): e00631–16.

Börlin, Marine, Pauline Venet, Olivier Claisse, Franck Salin, Jean-Luc Legras, and Isabelle Masneuf-Pomarede. "Cellar-Associated Saccharomyces Cerevisiae Population Structure Revealed High-Level Diversity and Perennial Persistence at Sauternes Wine Estates." *Applied and Environmental Microbiology* 82, no. 10 (2016): 2909–18.

Borneman, Anthony R., Angus H. Forgan, Radka Kolouchova, James A. Fraser, and Simon A. Schmidt. "Whole Genome Comparison Reveals High Levels of Inbreeding and Strain Redundancy Across the Spectrum of Commercial Wine Strains of Saccharomyces Cerevisiae." *G3: Genes | Genomes | Genetics* 6, no. 4 (2016): 957–71.

Bramley, R. G. V., and P. S. Gardiner. "Underpinning Terroir with Data: A Quantitative Analysis of Biophysical Variation in the Margaret River Region of Western Australia." *Australian Journal of Grape and Wine Research* 27, no. 4 (2021): 420–30.

Castriota-Scanderbeg, Alessandro, Gisela E. Hagberg, Antonio Cerasa, Giorgia Committeri, Gaspare Galati, Fabiana Patria, Sabrina Pitzalis, Carlo Caltagirone, and Richard Frackowiak. "The Appreciation of Wine by Sommeliers: A Functional Magnetic Resonance Study of Sensory Integration." *NeuroImage* 25, no. 2 (2005): 570–78.

Crabtree, Herbert Grace. "The Carbohydrate Metabolism of Certain Pathological Overgrowths." *Biochemical Journal* 22, no. 5 (1928): 1289–98.

Cross, Robin, Andrew J. Plantinga, and Robert N. Stavins. "What Is the Value of Terroir?" *American Economic Review* 101, no. 3 (2011): 152–56.

Curry, Andrew. "How Ancient People Fell in Love with Bread, Beer and Other Carbs." *Nature* 594, no. 7864 (2021): 488–91.

Dal Bello, Federica, Cristina Lamberti, Marzia Giribaldi, Cristiano Garino, Monica Locatelli, Daniela Gastaldi, Claudio Medana, Laura Cavallarin, Marco Arlorio, and Maria Gabriella Giuffrida. "Multi-Target Detection of Egg-White and Pig Gelatin Fining Agents in Nebbiolo-Based Aged Red Wine by Means of NanoHPLC-HRMS." *Food Chemistry* 345 (2021): 128822.

Darack, Ed. "Weaponizing Weather: The Top Secret History of Weather Modification." *Weatherwise* 72, no. 2 (2019): 24–31.

Dashko, Sofia, Nerve Zhou, Concetta Compagno, and Jure Piškur. "Why, When, and How Did Yeast Evolve Alcoholic Fermentation?" *FEMS Yeast Research* 14, no. 6 (2014): 826–32.

Davitashvili, Teimuraz, Inga Samkharadze, Lika Megreladze, and Ramaz Kvatadze. "Using Modern Technology to Protect Vineyards from Hail amid Climate Change." *E3S Web of Conferences* 234 (2021): 00034.

Day, M. P., D. Espinase Nandorfy, M. Z. Bekker, K. A. Bindon, M. Solomon, P. A. Smith, and S. A. Schmidt. "Aeration of *Vitis Vinifera* Shiraz Fermentation and Its Effect on Wine Chemical Composition and Sensory Attributes." *Australian Journal of Grape and Wine Research* 27, no. 3 (2021): 360–77.

Dessens, J., J. L. Sánchez, C. Berthet, L. Hermida, and A. Merino. "Hail Prevention by Ground-Based Silver Iodide Generators: Results of Historical and Modern Field Projects." *Atmospheric Research* 170 (2016): 98–111.

Di Castelnuovo, Augusto, Simona Costanzo, Marialaura Bonaccio, Patrick McElduff, Allan Linneberg, Veikko Salomaa, Satu Männistö, et al. "Alcohol Intake and Total Mortality in 142 960 Individuals from the MORGAM Project: A Population-Based Study." *Addiction* 117, no. 2 (2021): 312–25.

Doyon, Gilles, Alain Clément, Sabine Ribéreau, and Gérald Morin. "Canadian Bag-in-Box Wine Under Distribution Channel Abuse: Material Fatigue, Flexing Simulation and Total Closure/Spout Leakage Investigation." *Packaging Technology and Science* 18, no. 2 (2005): 97–106.

Dunn, Miriam, Mark D. A. Rounsevell, Fredrik Boberg, Elizabeth Clarke, Jens Christensen, and Marianne S. Madsen. "The Future Potential for Wine Production in Scotland Under High-End Climate Change." *Regional Environmental Change* 19, no. 3 (2019): 723–32.

Eisinger, Josef. "Lead and Wine: Eberhard Gockel and the Colica Pictonum." *Medical History* 26 (1982): 279–302.

Frank, Mitch. "Mixed Case: Turning Wine into Water and Creating Fear out of Nothing." *Wine Spectator*, April 7, 2015. https://www.winespectator.com/articles/turning-wine-into-water-and-creating-fear-out-of-nothing-51451.

Franklin, W. S. "Weather Control." *Science* 14, no. 352 (1901): 496–97.

Frost, Ram, Ileana Quiñones, Maria Veldhuizen, Jose-Iñaki Alava, Dana Small, and Manuel Carreiras. "What Can the Brain Teach Us About Winemaking? An fMRI Study of Alcohol Level Preferences." *PLOS ONE* 10, no. 3 (2015): e0119220.

Gaby, Jessica M., Alyssa J. Bakke, Allison N. Baker, Helene Hopfer, and John E. Hayes. "Individual Differences in Thresholds and Consumer Preferences for Rotundone Added to Red Wine." *Nutrients* 12, no. 9 (2020): E2522.

García-Estévez, I., M. T. Escribano-Bailón, J. C. Rivas-Gonzalo, and C. Alcalde-Eon. "Effect of the Type of Oak Barrels Employed During Ageing on the Ellagitannin Profile of Wines." *Australian Journal of Grape and Wine Research* 23, no. 3 (2017): 334–41.

Gawel, R., A. Schulkin, P. A. Smith, D. Espinase, and J. M. McRae. "Effect of Dissolved Carbon Dioxide on the Sensory Properties of Still White and Red Wines." *Australian Journal of Grape and Wine Research* 26, no. 2 (2020): 172–79.

Geffroy, Olivier, Tracey Siebert, Anthony Silvano, and Markus Herderich. "Impact of Winemaking Techniques on Classical Enological Parameters and Rotundone in Red Wine at the Laboratory Scale." *American Journal of Enology and Viticulture* 68, no. 1 (2017): 141–46.

Goddard, Matthew R. "Microbiology: Mixing Wine, Chocolate, and Coffee." *Current Biology* 26, no. 7 (2016): R275–77.

Goldstein, Nora. "Solid Waste Company Equipped for Increased Organics." *BioCycle*, September 14, 2020. https://www.biocycle.net/solid-waste-comp any-equipped-for-increased-organics/.

González-Centeno, María Reyes, Sophie Tempère, Pierre-Louis Teissedre, and Kleopatra Chira. "Use of Alimentary Film for Selective Sorption of Haloanisoles from Contaminated Red Wine." *Food Chemistry* 350 (2021): 128364.

Grotheer, Paul, Maurice Marshall, and Amy Simonne. "Sulfites: Separating Fact from Fiction." University of Florida Extension, March 17, 2019.

Gunnison, A. F., and D. W. Jacobsen. "Sulfite Hypersensitivity. A Critical Review." *CRC Critical Reviews in Toxicology* 17, no. 3 (1987): 185–214.

Hall, A., and G. V. Jones. "Spatial Analysis of Climate in Winegrape-Growing Regions in Australia." *Australian Journal of Grape and Wine Research* 16, no. 3 (2010): 389–404.

Hansel, Boris, Ronan Roussel, Vincent Diguet, Amandine Deplaude, M. John Chapman, and Eric Bruckert. "Relationships Between Consumption of Alcoholic Beverages and Healthy Foods: The French Supermarket Cohort of 196,000 Subjects." *European Journal of Preventive Cardiology* 22, no. 2 (2015): 215–22.

Haraway, Donna. *Staying with the Trouble*. Duke University Press, 2016.

Harper, Kristine C. "Climate Control: United States Weather Modification in the Cold War and Beyond." *Endeavour* 32, no. 1 (2008): 20–26.

Hjelmeland, Anna K., Thomas S. Collins, Joshua L. Miles, Philip L. Wylie, Alyson E. Mitchell, and Susan E. Ebeler. "High-Throughput, Sub Ng/ L Analysis of Haloanisoles in Wines Using HS-SPME with GC-Triple Quadrupole MS." *American Journal of Enology and Viticulture* 63, no. 4 (2012): 494–99.

Hohmann-Marriott, Martin F., and Robert E. Blankenship. "Evolution of Photosynthesis." *Annual Review of Plant Biology* 62 (2011): 515–48.

Jackowetz, Nick, Erhu Li, and Ramón Mira de Orduña. "Sulphur Dioxide Content of Wines: The Role of Winemaking and Carbonyl Compounds." *Appellation Cornell*, no. 3 (2011): 7.

Jarosz, Daniel F., Alex K. Lancaster, Jessica C. S. Brown, and Susan Lindquist. "An Evolutionarily Conserved Prion-like Element Converts Wild Fungi from Metabolic Specialists to Generalists." *Cell* 158, no. 5 (2014): 1072–82.

Jones, Gregory V. "Spatial Variability in Climate, Phenology, and Fruit Composition Across a Reference Vineyard Network in Southern Oregon." *E3S Web of Conferences* 50 (2018): 01018.

Jones, Gregory V. "Vintage Ratings: Applications of a Ranking Procedure to Facilitate a Better Understanding of Climate's Role in Wine Quality." *IVES Technical Reviews, Vine and Wine*, May 20, 2020.

Kampfer, Kristina, Alexander Leischnig, Björn Sven Ivens, and Charles Spence. "Touch-Flavor Transference: Assessing the Effect of Packaging Weight on

Gustatory Evaluations, Desire for Food and Beverages, and Willingness to Pay." *PLOS ONE* 12, no. 10 (2017): e0186121.

Keating, Grant Bartlett. "An Empirical Analysis of the Effect of Sub-Divisions of American Viticultural Areas on Wine Prices: A Hedonic Study of Napa Valley." *Journal of Wine Economics* 15, no. 3 (2020): 312–29.

Keller, Evelyn Fox. "Language and Ideology in Evolutionary Theory: Reading Cultural Norms into Natural Law." In *The Boundaries of Humanity*, edited by James J. Sheehan and Morton Sosna, 85–102. Berkeley: University of California Press, 1991.

King, Ellena S., Randall L. Dunn, and Hildegarde Heymann. "The Influence of Alcohol on the Sensory Perception of Red Wines." *Food Quality and Preference* 28, no. 1 (2013): 235–43.

Knight, Sarah, Steffen Klaere, Bruno Fedrizzi, and Matthew R. Goddard. "Regional Microbial Signatures Positively Correlate with Differential Wine Phenotypes: Evidence for a Microbial Aspect to Terroir." *Scientific Reports* Durham NC. 5 (2015): art. 14233.

Krymchantowski, Abouch Valenty, and Carla da Cunha Jevoux. "Wine and Headache." *Headache: The Journal of Head and Face Pain* 54, no. 6 (2014): 967–75.

"The Medical Council on Alcohol: MCA History." Accessed January 20, 2022. https://www.m-c-a.org.uk/Home/mca_history.

Kudo, Masayoshi, Paola Vagnoli, and Linda F. Bisson. "Imbalance of PH and Potassium Concentration as a Cause of Stuck Fermentations." *American Journal of Enology and Viticulture* 49, no. 3 (1998): 295–301.

Laaninen, Tarja. "Alcohol Labelling." European Parliament Research Service, September 2021. https://www.europarl.europa.eu/RegData/etudes/BRIE/2021/690563/EPRS_BRI(2021)690563_EN.pdf.

Larson, Greger, and Dorian Q. Fuller. "The Evolution of Animal Domestication." *Annual Review of Ecology, Evolution, and Systematics* 45, no. 1 (2014): 115–36.

Latour, Bruno. *The Pasteurization of France*. Translated by Alan Sheridan and John Law. Cambridge, MA: Harvard University Press, 1993.

Le Fur, Y., and L. Gautier. "De la minéralité dans les rosés?" *Revue française d'œnologie* 260 (2013): 40–43.

Leeuwen, Cornelis van, Philippe Friant, Xavier Choné, Olivier Tregoat, Stephanos Koundouras, and Denis Dubourdieu. "Influence of Climate, Soil, and Cultivar on Terroir." *American Journal of Enology and Viticulture* 55, no. 3 (2004): 207–17.

López-Rituerto, Eva, Francesco Savorani, Alberto Avenoza, Jesús H. Busto, Jesús M. Peregrina, and Søren Balling Engelsen. "Investigations of La Rioja Terroir for Wine Production Using 1H NMR Metabolomics." *Journal of Agricultural and Food Chemistry* 60, no. 13 (2012): 3452–61.

Lück, E. "Sulfur Dioxide." In *Antimicrobial Food Additives*. Berlin: Springer-Verlag, 1997.

Lyu, Xiaotong, Leandro Dias Araujo, Siew-Young Quek, and Paul A. Kilmartin. "Effects of Antioxidant and Elemental Sulfur Additions at Crushing on Aroma Profiles of Pinot Gris, Chardonnay and Sauvignon Blanc Wines." *Food Chemistry* 346 (June 1, 2021): 128914.

Maltman, Alex. "Minerality in Wine: A Geological Perspective." *Journal of Wine Research* 24, no. 3 (2013): 169–81.

Marchal, Axel, Philippe Marullo, Virginie Moine, and Denis Dubourdieu. "Influence of Yeast Macromolecules on Sweetness in Dry Wines: Role of the Saccharomyces Cerevisiae Protein Hsp12." *Journal of Agricultural and Food Chemistry* 59, no. 5 (2011): 2004–10.

Marchal, Axel, Pierre Waffo-Téguo, Eric Génin, Jean-Michel Mérillon, and Denis Dubourdieu. "Identification of New Natural Sweet Compounds in Wine Using Centrifugal Partition Chromatography-Gustatometry and Fourier Transform Mass Spectrometry." *Analytical Chemistry* 83, no. 24 (2011): 9629–37.

Martin, Laura B., Julie A. Nordlee, and Steve L. Taylor. "Sulfite Residues in Restaurant Salads." *Journal of Food Protection* 49, no. 2 (February 1986): 126–29.

Martiniuk, Jonathan T., Braydon Pacheco, Gordon Russell, Stephanie Tong, Ian Backstrom, and Vivien Measday. "Impact of Commercial Strain Use on Saccharomyces Cerevisiae Population Structure and Dynamics in Pinot Noir Vineyards and Spontaneous Fermentations of a Canadian Winery." *PLOS ONE* 11, no. 8 (2016): e0160259.

Mc Intyre, G. N., W. M. Kliewer, and L. A. Lider. "Some Limitations of the Degree Day System as Used in Viticulture in California." *American Journal of Enology and Viticulture* 38, no. 2 (January 1, 1987): 128–32.

McGovern, Patrick, Mindia Jalabadze, Stephen Batiuk, Michael P. Callahan, Karen E. Smith, Gretchen R. Hall, Eliso Kvavadze, et al. "Early Neolithic Wine of Georgia in the South Caucasus." *Proceedings of the National Academy of Sciences* 114, no. 48 (2017): 201714728.

National Research Council. *Critical Issues in Weather Modification Research*. Washington, DC: National Academies Press, 2003. https://doi.org/10.17226/10829.

Nietzsche, Friedrich. "Truth and Falsity in an Extra-Moral Sense." *ETC: A Review of General Semantics* 49, no. 1 (1992): 58–72.

Pabst, Evelyn, Armando Maria Corsi, Riccardo Vecchio, Azzurra Annunziata, and Simone Mueller Loose. "Consumers' Reactions to Nutrition and Ingredient Labelling for Wine – A Cross-Country Discrete Choice Experiment." *Appetite* 156 (2021): 104843.

Paparazzo, Ernesto. "Philosophy and Science in the Elder Pliny's Naturalis Historia." In *Pliny the Elder: Themes and Contexts*, edited by Roy Gibson and Ruth Morello, 89–112. Leiden: Brill, 2011.

Parr, Wendy V., Alex J. Maltman, Sally Easton, and Jordi Ballester. "Minerality in Wine: Towards the Reality Behind the Myths." *Beverages* 4, no. 4 (2018): 77.

Parr, Wendy V., Dominique Valentin, Jason Breitmeyer, Dominique Peyron, Philippe Darriet, Robert Sherlock, Brett Robinson, Claire Grose, and Jordi Ballester. "Perceived Minerality in Sauvignon Blanc Wine: Chemical Reality or Cultural Construct?" *Food Research International* 87 (2016): 168–79.

Pastore, Chiara, Marisa Fontana, Stefano Raimondi, Paola Ruffa, Ilaria Filippetti, and Anna Schneider. "Genetic Characterization of Grapevine Varieties from Emilia-Romagna (Northern Italy) Discloses Unexplored Genetic Resources." *American Journal of Enology and Viticulture* 71, no. 4 (2020): 334–43.

Paul, Harry W. *Science, Vine and Wine in Modern France.* Cambridge: Cambridge University Press, 2002.

Paxson, Heather. "Post-Pasteurian Cultures: The Microbiopolitics of Raw-Milk Cheese in the United States." *Cultural Anthropology* 23, no. 1 (2008): 15–47.

Perdue, Andy. "Making Wine for High Altitudes." *Wine Business Monthly*, July 2014. https://www.winebusiness.com/wbm/?go=getArticle&dataId= 134437.

Piqueras-Fiszman, Betina, and Charles Spence. "The Weight of the Bottle as a Possible Extrinsic Cue with Which to Estimate the Price (and Quality) of the Wine? Observed Correlations." *Food Quality and Preference* 25, no. 1 (2012): 41–45.

Piškur, Jure, Elżbieta Rozpędowska, Silvia Polakova, Annamaria Merico, and Concetta Compagno. "How Did Saccharomyces Evolve to Become a Good Brewer?" *Trends in Genetics* 22, no. 4 (2006): 183–86.

Rankine, B. C. "Developments in Malo-Lactic Fermentation of Australian Red Table Wines." *American Journal of Enology and Viticulture* 28, no. 1 (1977): 27–33.

Research and Markets. "$2+ Billion Worldwide Tartaric Acid Industry to 2026—Impact Analysis of COVID-19." November 8, 2021. https://www.prn ewswire.com/news-releases/2-billion-worldwide-tartaric-acid-industry- to-2026---impact-analysis-of-covid-19-301418817.html.

Revelette, Matthew R., Jennifer A. Barak, and James A. Kennedy. "High-Performance Liquid Chromatography Determination of Red Wine Tannin Stickiness." *Journal of Agricultural and Food Chemistry* 62, no. 28 (2014): 6626–31.

Rinaldi, Alessandra, Angelita Gambuti, and Luigi Moio. "Precipitation of Salivary Proteins After the Interaction with Wine: The Effect of Ethanol, pH, Fructose, and Mannoproteins." *Journal of Food Science* 77, no. 4 (2012): C485–90.

Robinson, Jancis, Julia Hardin, and José Vouillamoz. *Wine Grapes: A Complete Guide to 1,368 Vine Varieties, Including Their Origins and Flavours.* London: Penguin UK, 2012.

Roullier-Gall, Chloé, Daniel Hemmler, Michael Gonsior, Yan Li, Maria Nikolantonaki, Alissa Aron, Christian Coelho, Régis D. Gougeon, and Philippe Schmitt-Kopplin. "Sulfites and the Wine Metabolome." *Food Chemistry* 237 (December 15, 2017): 106–13.

Roullier-Gall, Chloé, Marianna Lucio, Laurence Noret, Philippe Schmitt-Kopplin, and Régis D. Gougeon. "How Subtle Is the 'Terroir' Effect? Chemistry-Related Signatures of Two 'Climats de Bourgogne.'" *PloS One* 9, no. 5 (2014): e97615.

Royle, John Forbes. *On the Production of Isinglass Along the Coasts of India: With a Notice of Its Fisheries.* London: Allen, 1842.

Santé Publique France. "L'alcool pour comprendre." 2021. https://www.sant epubliquefrance.fr/import/l-alcool-pour-comprendre.

Schad, Susanne G., Jiri Trcka, Stefan Vieths, Stephan Scheurer, Amedeo Conti, Eva-B. Brocker, and Axel Trautmann. "Wine Anaphylaxis in a German Patient: IgE-Mediated Allergy Against a Lipid Transfer Protein of Grapes." *International Archives of Allergy and Immunology* 136, no. 2 (2005): 159–64.

Schamel, Günter, and Kym Anderson. "Wine Quality and Varietal, Regional and Winery Reputations: Hedonic Prices for Australia and New Zealand." *Economic Record* 79, no. 246 (2003): 357–69.

Schmidt, Liane, Vasilisa Skvortsova, Claus Kullen, Bernd Weber, and Hilke Plassmann. "How Context Alters Value: The Brain's Valuation and Affective Regulation System Link Price Cues to Experienced Taste Pleasantness." *Scientific Reports* 7, no. 1 (2017): 8098.

Schöbel, Nicole, Debbie Radtke, Jessica Kyereme, Nadine Wollmann, Annika Cichy, Katja Obst, Kerstin Kallweit, et al. "Astringency Is a Trigeminal Sensation That Involves the Activation of G Protein–Coupled Signaling by Phenolic Compounds." *Chemical Senses* 39, no. 6 (2014): 471–87.

Schöbel, Nicole, Debbie Radtke, Jessica Kyereme, Nadine Wollmann, Annika Cichy, Katja Obst, Kerstin Kallweit, et al. "Astringency Is a Trigeminal Sensation That Involves the Activation of G Protein–Coupled Signaling by Phenolic Compounds." *Chemical Senses* 39, no. 6 (2014): 471–87.

Schultze, Steven R., Paolo Sabbatini, and Lifeng Luo. "Effects of a Warming Trend on Cool Climate Viticulture in Michigan, USA." *SpringerPlus* 5, no. 1 (2016): 1119.

Secretary of the General Assembly. "OIV Process for the Clonal Selection of Vines," 2017. https://www.oiv.int/public/medias/5382/oiv-viti-564a-2017-en.pdf.

Shimizu, Hideaki, Fumikazu Akamatsu, Aya Kamada, Kazuya Koyama, Kazuhiro Iwashita, and Nami Goto-Yamamoto. "Variation in the Mineral Composition of Wine Produced Using Different Winemaking Techniques." *Journal of Bioscience and Bioengineering* 130, no. 2 (2020): 166–72.

Simpson, James. *Creating Wine: The Emergence of a World Industry, 1840–1914.* Princeton: Princeton University Press, 2011.

Simpson, Robert F., and Mark A. Sefton. "Origin and Fate of 2,4,6-Trichloroanisole in Cork Bark and Wine Corks." *Australian Journal of Grape and Wine Research* 13, no. 2 (2007): 106–16.

Starkenmann, Christian, Charles Jean-Francois Chappuis, Yvan Niclass, and Pascale Deneulin. "Identification of Hydrogen Disulfanes and Hydrogen Trisulfanes in H_2S Bottle, in Flint, and in Dry Mineral White Wine." *Journal of Agricultural and Food Chemistry* 64, no. 47 (2016): 9033–40.

Steensels, Jan, Brigida Gallone, Karin Voordeckers, and Kevin J. Verstrepen. "Domestication of Industrial Microbes." *Current Biology* 29, no. 10 (2019): R381–93.

Streletskaya, Nadia A., Jura Liaukonyte, and Harry M. Kaiser. "Absence Labels: How Does Information About Production Practices Impact Consumer Demand?" *PLOS ONE* 14, no. 6 (2019): e0217934.

Sun, Marjorie. "Salad, House Dressing, but Hold the Sulfites." *Science* 226, no. 4674 (1984): 520–21.

Szymanski, Erika Amethyst. "Through the Grapevine: In Search of a Rhetoric of Industry-Oriented Science Communication." PhD thesis, University of Otago, 2017.

Thatch, Liz, and Angelo Camillo. "A Snapshot of the American Wine Consumer in 2018." *Wine Business*, December 10, 2018.

Tomasi, Diego, Gregory V. Jones, Mirella Giust, Lorenzo Lovat, and Federica Gaiotti. "Grapevine Phenology and Climate Change: Relationships and Trends in the Veneto Region of Italy for 1964–2009." *American Journal of Enology and Viticulture* 62, no. 3 (2011): 329–39.

Varela, C., P. R. Dry, D. R. Kutyna, I. L. Francis, P. A. Henschke, C. D. Curtin, and P. J. Chambers. "Strategies for Reducing Alcohol Concentration in Wine." *Australian Journal of Grape and Wine Research* 21, no. S1 (2015): 670–79.

Vasiljevic, Milica, Dominique-Laurent Couturier, and Theresa M. Marteau. "Impact of Low Alcohol Verbal Descriptors on Perceived Strength: An Experimental Study." *British Journal of Health Psychology* 23, no. 1 (2018): 38–67.

Vezzulli, Silvia, Lorena Leonardelli, Umberto Malossini, Marco Stefanini, Riccardo Velasco, and Claudio Moser. "Pinot Blanc and Pinot Gris Arose as Independent Somatic Mutations of Pinot Noir." *Journal of Experimental Botany*, October 23, 2012, ers290.

Wang, Qian (Janice), and Charles Spence. "Assessing the Influence of Music on Wine Perception Among Wine Professionals." *Food Science and Nutrition* 6, no. 2 (2017): 295–301.

Williamson, P. O., S. Mueller-Loose, L. Lockshin, and I. L. Francis. "More Hawthorn and Less Dried Longan: The Role of Information and Taste on Red Wine Consumer Preferences in China." *Australian Journal of Grape and Wine Research* 24, no. 1 (2018): 113–24.

Winkler, A. J. "The Utilization of Sulfur Dioxide in the Marketing of Grapes." *Hilgardia* 1, no. 6 (June 1925): 107–31.

Zhang, Nansen, Andrew Hoadley, Jim Patel, Seng Lim, and Chaoen Li. "Sustainable Options for the Utilization of Solid Residues from Wine Production." *Waste Management* 60, supplement C, Special Thematic Issue: Urban Mining and Circular Economy (2017): 173–83.

Index

For the benefit of digital users, indexed terms that span two pages (e.g., 52–53) may, on occasion, appear on only one of those pages.

Figures and boxes are indicated by *f* and *b* following the page number